SOIL COMPACTION AND REGENERATION

*Proceedings of the Workshop on Soil Compaction:
Consequences and Structural Regeneration Processes / Avignon
17-18 September 1985*

SOIL COMPACTION AND REGENERATION

Edited by
G.MONNIER
INRA, Montfavet

M.J.GOSS
Rothamsted Experimental Station, Harpenden

Published for the Commission of the European Communities by
A.A.BALKEMA / ROTTERDAM / BOSTON / 1987

Published for the Commission of the European Communities,
Directorate-General Telecommunications, Information Industries & Innovation, Luxembourg

LEGAL NOTICE
Neither the Commission of the European Communities, nor any person acting on behalf of
the Commission is responsible for the use which might be made of the following information.

EUR 10396
ISBN 90 6191 780 8
© ECSC, EEC, EAEC, Brussels and Luxembourg, 1987

Published by A.A.Balkema, P.O.Box 1675, 3000 BR Rotterdam, Netherlands
Printed in the Netherlands

Table of contents

V

General introduction

Since the establishment of the European Economic Community, changes in the technology of crop production in member countries have increased the risks and hence potential importance of soil compaction. Firstly, intensive management and the tendency towards monoculture have tended to limit the time available for tillage and harvest so constraining farmers to operate more often in unfavourable, wet conditions. The increase in power and weight of tractors, harvesting machinery and trailers has aggravated the situation. Extra power allows equipment to move on wet soils so increasing the likelihood of damage. Extra axle weight requires greater bearing capacity and hence soils are more likely to be damaged.

The consequences of these developments in agriculture for crop production and soil conservation can be serious. For example, compaction may reduce hydraulic conductivity and hence restrict water movement through the soil leading to waterlogging, breakdown of drainage systems and a lowering of the effective use of irrigation water.

Similarly more compact soil has reduced gas diffusivity which limits biological activity in soil and hence adversely affects the cycling of nutrients such as nitrogen and sulphur.

In compacted soil, root systems are frequently shallow and sparsely branched leading to a greater susceptibility to drought and a reduced efficiency of fertilizer utilization.

To avoid these possible effects the practical response has been an increase in energy requirements for tillage to overcome the damage to soil.

Until the last few years the consequences of compaction for crop yields have been partly hidden by the rapid progress in crop protection, soil fertility and plant breeding. However, annual improvements in yield by these means have become progressively less so that limitations to yield due to other factors have become more obvious. Now the increased production costs due to the lowered efficiency of utilising inputs is becoming clearer. This new situation has encouraged new programmes of research in many European Countries, but we have insufficient knowledge to solve all the problems. There is need of accurate methods for describing and predicting all the changes to soil resulting from compaction. We must be able to estimate the natural regeneration of structure following compaction. In this way, it should be possible to make recommendations to prevent compaction and alleviate or correct soil damage under different cropping systems in the main climatic and soil zones.

The aim of this workshop was to discuss and compare the methods available for studying compaction and structural regeneration and, starting from the current state of knowledge, define the main questions for research and identify profitable links where collaboration would improve research capabilities.

Each session of the workshop was centred on a number of invited papers but some shorter contributions were also presented and these are included in the Proceedings.

The first session was devoted to the physical criteria for describing soil compaction and the efficacy of various techniques for assessing them. The approaches covered the direct measurement of the state of the soil and its evolution by compaction. Also included were those techniques, such as those to measure fluid transport characteristics, that seek to quantify soil properties that may be modified by compaction. Since a single measurement can provide only a simplified way of describing, albeit quantitatively, a complex reality the second session considered micro- and macro-morphological approaches to the study of soil structure evolution during compaction.

In the third session consideration was given to the application of modelling to describe and understand the processes involved in structural regeneration. Particular attention was given to factors that affect the susceptibility of soils to compaction and the soil's tendency to recover from damage.

The influence of soil compaction on crop growth and yield together with the consequences of tillage in improving some compacted soils, but making others more susceptible to compaction, was discussed in the next session.

Plant roots were the main topic in the fifth session. Condideration was given to the fact that not only are roots affected by soil physical conditions but they themselves can and do modify soil properties.

As well as providing a record of the workshop the aim of this Proceeding is to stimulate further discussion of the topics. A synthesis of the discussions that took place during the sessions and in the closing session are also included.

We are happy to thank all the contributors and participants to the workshop and the European Commission which sponsored this meeting.

G. MONNIER

L'analyse de la porosité: Application à l'étude du compactage des sols

J.Guérif

Institut National de la Recherche Agronomique, Centre de Recherches Agronomiques d'Avignon, Station de Science du Sol, Montfavet, France

RESUME

L'étude du compactage et de ses conséquences soulève de nombreux problèmes méthodologiques et notamment :
- Comment exprimer l'effet du compactage sur le sol ?
- Comment modéliser le processus de compactage des couches de surface sans satisfaire aux conditions d'homogénéité, d'isotropie et de continuité ?
- Comment évaluer en terme de risques pour la plante les effets du compactage ?

L'auteur propose d'utiliser comme état de référence la masse volumique résultante de l'assemblage des constituants évaluée sur des agrégats de 2-3 mm de diamètre.

Cette partition de l'espace poral en textural (intra-agrégats) et structural (inter-agrégats) permet une bonne interprétation du niveau de masse volumique, quelle que soit la teneur en eau.

Une analyse du processus de compactage, en distinguant des comportements mécaniques texturaux et structuraux, est proposée.

Des exemples de l'application de l'analyse des systèmes de porosité à l'étude des conséquences du compactage sont présentés et plus particulièrement l'interaction entre espace poral structural et transferts d'eau et de gaz (perméabilité, diffusivité) ainsi que les propriétés mécaniques résultantes.

ABSTRACT : **SOIL COMPACTION : PORE SPACE ANALYSIS EFFICIENCY.**

Several difficulties arise when studying soil compaction and its consequences on crop growth :
- How to express compaction responses in the soil ?
- How to predict soil compaction under wheel without assuming homogeneity, continuity and isotropy of top soil layers ?
- How to assess the real risks of crops consecutive to soil compaction ?

The author proposes to use, as a reference state, the bulk density due only to the packing of soil constituents, which may be measured on aggregates of 2-3 mm diameter.

This leads to a partition of pore space into "textural" (intra-aggregates) and "structural" (inter-aggregates) pore spaces, which allows a good interpretation of bulk density at any water content.

An analysis of compaction process, distinguishing textural and structural mechanical behaviours is proposed.

Examples of using this pore space partition for the study of the consequences of compaction are given. The interaction between structural pore space and water and gas transfer (permeability, diffusivity) as well as mechanical properties (soil strength) are emphasized.

INTRODUCTION

Les difficultés d'étudier le compactage des sols agricoles, et surtout de l'interpréter en terme de risque pour les cultures en cours et à venir, sont d'ordre différents et concernent :

- **L'évaluation des modifications de l'état physique** de la couche de sol tassée, le choix des variables à analyser et donc le choix des mesures physiques à mettre en œuvre. Ainsi, l'augmentation de compacité, appréciée en bilan de volume, est une variable qui décrit insuffisamment les modifications de l'espace poral (formes des pores, organisation, connexité, ...) qui sont à l'origine des modifications des propriétés de transfert (gaz, liquide, chaleur) ou des propriétés mécaniques (résistance à la pénétration, résistance à la rupture) susceptibles d'affecter le fonctionnement du système racinaire (fonction puit, implantation) directement ou indirectement par la qualité du travail du sol postérieur à la phase de compactage.

- **La modélisation des phénomènes de compactage** (nécessaire pour l'analyse des mécanismes, comme pour la prévision de la dégradation des sols). Les couches de surface des sols agricoles ne répondent que rarement aux hypothèses d'homogénéité, de continuité, et d'isotropie, généralement admises dans les modèles existants d'une part, au niveau de l'évaluation de la transmission des contraintes en profondeur (on utilise généralement la formule de Soëhn) d'autre part, et dans une moindre mesure, au niveau de la relation contrainte déformation (Guérif, 1984).

- **L'évaluation du niveau de risque** qu'entrainent les modifications de l'état physique du sol par compactage et leurs conséquences pour le fonctionnement de la plante en place ou pour l'implantation de la future culture. En effet, cette évaluation doit être pondérée en fonction de la nature et de l'intensité des évènements climatiques probables. Ceux-ci peuvent, en effet, augmenter ou atténuer les risques, voire contribuer, à une régénération de l'état structural.

EXPRESSIONS PHYSIQUES DE LA COMPACITE

La compacité ou ses variations sont le plus souvent évaluées directement par des mesures combinées de masse et de volume ou indirectement par des mesures étalonnées, bien sûr, par rapport à des mesures de masse et de volume, telles les mesures de variation de cote ou les mesures de l'atténuation ou de la rétrodiffusion de rayonnement gamma (Soane, 1976 ; Soane and Henshal, 1979 ; Stengel, 1985 ; Raghavan et al., 1976).

Dans la majorité des cas, dans la littérature internationale, les résultats de ces mesures s'expriment donc sous forme de masse-volumique, rapport de la masse de solide au volume de l'échantillon de sol considéré.

Suivant les disciplines (Hydrologie, Génie Civil, Agronomie) et suivant les auteurs, la compacitée est exprimée par différentes combinaisons de bilan de volume.

$$\frac{\text{volume des pores}}{\text{volume total}} \quad : \text{porosité (porosity) } n \text{ \%}$$

$$\frac{\text{volume des pores}}{\text{volume de solide}} \quad : \text{indice des vides (void ratio) } \mathbf{e} \quad \text{Génie Civil}$$

$$\frac{\text{volume total}}{\text{volume de solide}} \quad : \text{volume spécifique (specific volume) } 1 + \mathbf{e}$$

Kurtay and Reece : Critical State theory

$$\frac{\text{volume de solide}}{\text{volume total}} \quad : \text{"packing density"} \quad \frac{1}{1 + \mathbf{e}} = D$$

Dexter et Tanner

LES ETATS DE REFERENCES

Dans la littérature

Quelle que soit l'expression de compacité choisie, cette variable ne permet pas, d'une manière satisfaisante, d'interpréter des variations de comportement (hydriques (transferts), ou mécaniques) d'un sol à un autre, ou entre deux états physiques d'un même sol.

De nombreux auteurs ont tenté de résoudre le problème en définissant un système de référence permettant de situer un état physique donné par rapport à un état physique standard considéré comme caractéristique du matériau.

Ainsi, pour relativiser l'intensité d'un tassement, on peut faire référence à un état de compacité obtenu par un protocole de compactage standard.

- Une charge et son mode d'application

- masse volumique au maximum Proctor ($\rho_{d\ max}$). Pigeon et Soane (1977) ont ainsi défini une compacité relative ("relative compaction") $\rho_d / \rho_{d\ max}$

- masse volumique obtenu après une compression statique uniaxiale confinée de 200 kPa (2 kg/cm²). Hakansson (Eriksson et al., 1974) a défini de la même manière un "degré de compacité" $\rho_d / \rho_{d\ standard}$).100 % (degree of compaction). Aucune caractéristique physique initiale de l'échantillon (notamment hydrique) n'est définie.

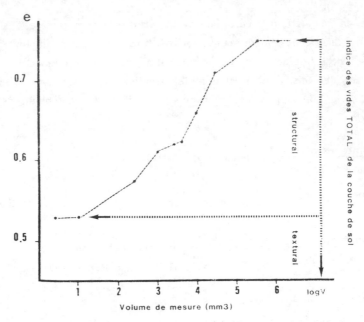

Fig. 1 Variations de l'indice des vides (**e**) d'une couche de sol en fonction du volume de mesure (V).
Variation of the top soil layer void ratio with the volume of soil upon which the measurement is done.

- Un test de compactage par incréments successifs de la charge.

L'état de référence est défini par l'asymptote de la courbe de compression (Dexter et Tanner, 1973).

Par ailleurs, pour relativiser le volume et la taille des pores, d'un sol à l'autre et d'un état physique à l'autre, on fait fréquemment appel à la notion de micro et de macroporosité. La limite arbitraire entre ces deux classes de pores est choisie pour une succion variable suivant les auteurs et en fonction des types de sol entre quelques dizaines et quelques centaines de mbar.

Ces systèmes de référence sont établis pour un état hydrique donné du matériau et ne permettent donc pas de caractériser l'état physique d'une couche de sol quelle que soit sa teneur en eau.

L'analyse du système de porosité

Fies (1971), Monnier et al. (1973), Stengel (1979), Fies et Stengel (1981) proposent un système de référence où l'hypothèse de base consiste à admettre qu'une fraction de l'espace poral résulte de l'assemblage des particules élémentaires constitutives du sol. Elle est déterminée essentiellement par les caractéristiques de ces particules (taille, forme, nature minéralogique, garniture ionique) et leur état d'hydratation. Cette fraction est appelée **porosité texturale.** Son complément dans la porosité totale est appelé **porosité structurale.** Cette dernière varie sous l'effet d'actions extérieures que subit le matériau : travail du sol et tassements mécaniques, actions climatiques et biologiques.

Le caractère opérationnel de cette analyse dépend de la capacité à mesurer l'une de ces fractions de l'espace poral.

L'analyse des relations entre le volume d'un échantillon et sa porosité nous a conduit à supposer que la porosité contenue dans un fragment de matériau dont le volume est suffisamment petit est essentiellement d'origine texturale (fig. 1). Et on a constaté qu'en mesurant la porosité d'agrégats obtenus par rupture et tamisage entre 2 et 3 mm, on obtenait une valeur, qui, pour un matériau donné :

- à l'état sec varie peu au cours du temps,

- est liée à la teneur en eau par une courbe de retrait-gonflement qu'on peut également considérer comme stable (fig. 2).

Par ailleurs, à l'état sec, cette porosité est étroitement corrélée à la constitution des matériaux (teneur en argile, teneur en matières organiques).

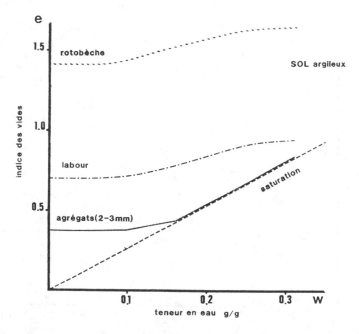

Fig. 2   Courbes de retrait de couches de surface d'un sol argileux travaillé et courbe de retrait textural évaluée sur des agrégats de 2-3 mm.
Shrinkage curves of a clayey soil, measured on tilled layers (rotary spade machine, and ploughed) and on aggregates (2-3 mm diameter).

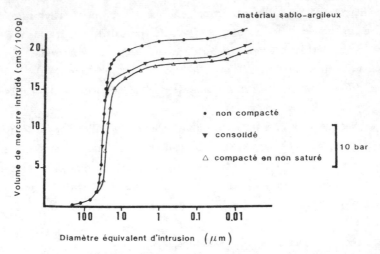

Fig. 3   Cou-bes d'intrusion de mercure : Effet du compactage sur l'espace poral textural d'un sol sablo-argileux.
Mercury intrusion curves : Effect of compaction on the "textural" pore space (sandy clay soil).

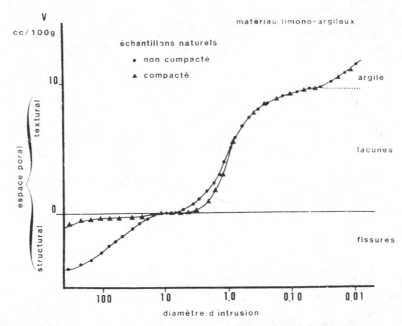

Fig. 4   Courbes d'intrusion de mercure : Effet du compactage sur l'espace poral textural d'un sol limono-argileux.
Mercury intrusion curves : Effect of compaction on the textural pore space (silty clay soil).

Cet ensemble de constatations permet d'assimiler la porosité d'agrégats à la porosité texturale.

La possibilité d'utiliser cette analyse pour l'étude des phénomènes de compactage et de leurs conséquences dépend de la stabilité de la porosité texturale vis-à-vis des actions de compactage, au moins dans la gamme de pressions rencontrées en conditions agricoles. Seul, l'assemblage des constituants des matériaux très sableux (fig. 3) semble affecté d'une manière significative par les actions mécaniques (Fies et Zimmer, 1982).

Il n'a pas été possible de mettre en évidence, par des procédés de compactage de laboratoire, des variations de l'assemblage des constituants pour les matériaux argilo-limoneux (fig. 4) (Fies et Guérif, non publié).

- La porosité structurale est seule, ou de façon très prédominante, affectée par l'action d'agents extérieurs (à une humidité donnée) (fig. 5). Elle est donc une meilleure caractéristique des effets de ces actions que la porosité totale. Vis-à-vis des actions anthropiques en particulier, et en adoptant un point de vue d'ingénieur, l'analyse de la porosité permet d'isoler la porosité sur laquelle on agit et celle qu'on ne peut pratiquement pas modifier.

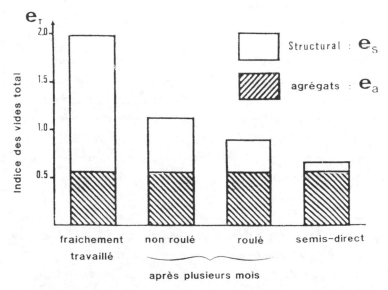

Fig. 5 Analyse du système de porosité à différents niveaux de tassement. Le sol est la capacité au champ.
Pore space partition at different state of compaction (just tilled, after several monthes : unwheeled and wheeled, direct-drilled).

- On peut par analyse de la porosité comparer les effets de ces actions à différentes teneurs en eau et dans différents matériaux, en tenant compte de l'état de gonflement du moment de la mesure. Cela est impossible à partir de mesures de la porosité totale ou de la division en macro et micro porosité, cette dernière n'étant valide qu'à une succion donnée.

L'ANALYSE DU SYSTEME DE POROSITE DANS L'ETUDE DES MECANISMES DE COMPACTAGE

Le comportement au compactage des couches de sols travaillées fait intervenir :

- les propriétés mécaniques résultant des contacts et frottements entre les éléments structuraux (mottes, agrégats, terre fine) constitutifs de la couche de sol.

- les propriétés mécaniques des éléments structuraux eux-mêmes.

Les tests de compactage, effectués en laboratoire, sur des massifs d'agrégats, sont à cet égard très éclairant.

Ainsi, Faure (1980) définit deux domaines hydriques de comportement au compactage.

- Un domaine où le compactage dépend du concassage et du réarrangement des agrégats,

- Un domaine où les agrégats deviennent déformables avec l'entrée en plasticité de la phase argileuse.

De même, Braun'ack et Dexter (1978) font intervenir dans leur modèle de compressibilité d'un massif, la résistance à l'écrasement propre aux agrégats. Collis-George et Lloyd (1978) estiment les contraintes de cisaillement d'un lit de semences à partir du frottement des agrégats entre eux et non du frottement entre constituants minéraux. Ce frottement entre agrégats diminue quand la teneur en eau augmente (Guérif, 1982).

On peut donc envisager d'interpréter les comportements au compactage des couches de sol travaillés et donc discontinues en distinguant :

- les caractéristiques mécaniques intrinsèques au matériau, résultantes de l'assemblage des éléments minéraux constitutifs et donc texturales.

- les caractéristiques mécaniques résultantes de l'assemblage des éléments structuraux (mottes et agrégats) et donc structurales.

On peut ainsi définir trois domaines hydriques de comportement au compactage (fig. 6) à partir des comportements mécaniques texturaux :

- $W < W_1$: l'amplitude du tassement est faible. Elle est fonction de la résultante de la pression appliquée et des frottements entre agrégats. Elle

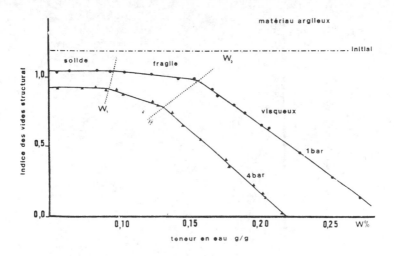

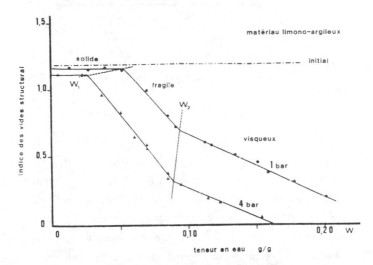

Fig. 6   Compactage de massifs d'agrégats : effet de la teneur en eau et de la pression appliquée sur l'indice des vides structuraux.
Compaction of aggregates'bed : effect of water-content and pressure on structural void ratio.

est indépendante du temps d'application. Les agrégats ont un comportement **SOLIDE**. Leur résistance à l'écrasement est grande par rapport aux contraintes qui se développent dans le massif.

- $W_1 \leqslant W \leqslant W_2$: à l'humidité W_1 la résistance à l'écrasement des agrégats est de l'ordre de grandeur des contraintes auxquelles ils sont soumis. L'augmen-

9

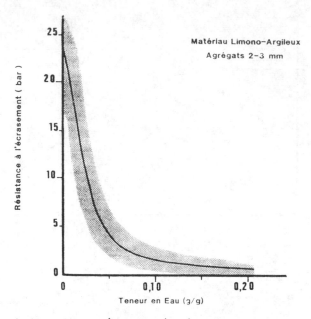

Fig. 7 Variations de la résistance à l'écrasement d'agrégats en fonction de leur teneur en eau.
"Textural" tensile-strength versus water content measured on aggregates (2-3 mm diameter).

tation de l'humidité s'accompagne d'une diminution rapide de la "cohésion texturale" (fig. 7). Les agrégats deviennent **FRAGILES.** Le tassement du massif résultera du réarrangement des petits morceaux issus de la rupture.

- $W \geqslant W_2$: dans cette gamme de teneur en eau, on ne peut plus considérer la déformation comme instantanée. Le comportement sous charge est alors **VISQUEUX.** L'amplitude de déformation est en effet fonction du temps d'application de la charge.

L'ANALYSE DU SYSTEME DE POROSITE DANS L'ETUDE DES CONSEQUENCES DU COMPACTAGE

Au niveau des propriétés mécaniques

L'augmentation de la résistance à la rupture d'un massif d'agrégats avec l'augmentation de sa compacité (fig. 8) par compactage dépend :

- de l'intensité des forces de liaison qui ont pu se rétablir entre éléments structuraux : le niveau de coalescence des agrégats dépend de l'interaction humidité-intensité de compactage.

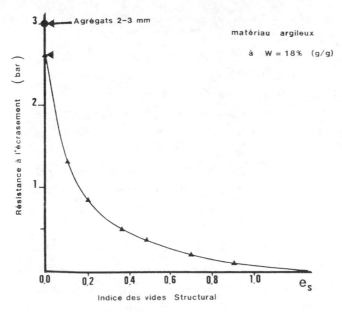

Fig. 8 Résistance à l'écrasement de massifs d'agrégats, compactés à une humidité donnée (ψ = 5 bar), en fonction de l'indice des vides structural.
"Structural" tensile strength at a given moisture content (ψ = 5 bar) versus structural void ratio.

- de la surface de contact entre éléments structuraux ; celle-ci est fortement corrélée à l'indice des vides structural.

<u>Au niveau des propriétés de transferts</u>

Une quantification de l'espace poral structural semble être une variable prometteuse dans la modélisation de certains phénomènes de transfert :

Ainsi, Fies a montré que le coefficient K de Darcy en saturé pouvait être ajusté de façon satisfaisante à une fonction ayant la forme de la relation de Kozeny en ne faisant intervenir que la porosité structurale (fig. 9). De même, Bruckler (à paraître) a montré que la diffusivité en phase gazeuse, évaluée avec le dispositif de Ball, pouvait être modélisée par la loi de Fick en faisant notamment intervenir une "porosité de transfert" pouvant s'interpréter comme l'espace poral structural.

CONCLUSION

Cette partition dans l'espace poral, en fonction de son origine soit texturale, soit structurale, semble prometteuse à la fois dans l'étude des mécanismes, mais aussi des conséquences du compactage des couches de surface des sols agricoles.

Elle répond à un souci de pouvoir faire référence à un état bien défini pour apprécier l'intensité du compactage mais aussi pour pouvoir comparer différents matériaux en différentes situations.

Elle devrait permettre une approche de la mécanique des milieux discontinus certes moins analytique, mais plus réaliste que celles qui existent actuellement.

En comptabilisant la part de l'espace poral affecté par le compactage, cette analyse du système de porosité devrait permettre d'analyser les conséquences des actions de compactage pour les propriétés mécaniques et les propriétés de transferts.

Il convient néanmoins de compléter cette évaluation aveugle du volume poral affecté par le compactage par une appréciation quantitative des modifications de ses caractéristiques morphologiques (cf. articles de Stengel et Pagliai).

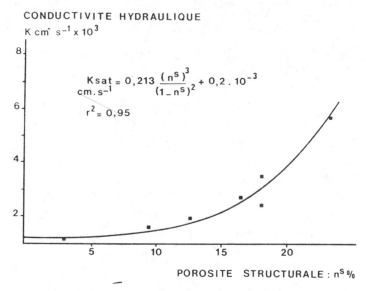

Fig. 9 Ajustement de la conductivité hydraulique à la porosité structurale par la formule de KOZENY (d'après J.C. Fies).
Curve fitting of the hydraulic conductivity with structural porosity by KOZENY's formula (from J.C. Fies).

REFERENCES

Ball, B. ; Harris, W. ; Burford J.R., 1981. A laboratory method to measure gaz diffusion and flow in soil and other porous material. J. of Soil Sci. 32: 323-333.

Dexter, A.R. and Tanner, D.W., 1973. The response of unsaturated soils to isotropic stress. J. of Soil Sci. 24: 491-502.

Eriksson, J. ; Hakansson, I. ; Danfors, B., 1974. The effect of soil compaction on soil structure and crop yields. Swedish Inst. of Agric. Engng, Bull. 354.

Faure, A., 1980-1981. A new conception of the plastic and liquid limits of clay. Soil Tillage Res. 1:97-105.

Fies, J.C., 1971. Recherche d'une interprétation texturale de la porosité des sols. Ann. agron. 22(6): 655-685.

Fies, J.C., 1982. Etude des écoulements en milieux saturés en relation avec la morphologie de l'espace poral du sol. Note interne, INRA-Sol, Montfavet.

Fies, J.C. and Zimmer, D., 1982. Etude expérimentale de modifications de l'assemblage textural d'un matériau sablo-argileux sous l'effet de pressions. Bull. du Groupe Français d'Humidimétrie neutronique, 12, 39-54.

Fies, J.C. and Stengel, P., 1981. Densité texturale de sols naturels. I-Méthode de mesure. II-Eléments d'interprétation. Agronomie 1(8): 651-658;659-666.

Guérif, J., 1982. Compactage d'un massif d'agrégats : effet de la teneur en eau et de la pression appliquée. Agronomie, 2(3), 287-294.

Guérif, J., 1984. The influence of water-gradient and structure anisotropy on soil compressibility. J. Agric. Engng 29:367-374.

Kurtay, T. and Reece, A.R., 1970. Plasticity theory and critial state soil mechanics. J. Terramechanics 7:23-56.

Pidgeon, J.D. and Soane, B.D., 1977. Effects of tillage and direct drilling on soil properties during the growing season in a long-term barley monoculture system. J. Agric. Sci. Camb., 88:431-442.

Raghavan, G.S.V. ; McKyes, E. ; Chassi, M., 1976. Soil compaction patterns caused by off-road vehicles in Eastern Canadian agricultural soils. J. Terramechanics, 13:107-115.

Soane, B.D., 1976. Gamma ray transmission systems for the in-situ measurement of soil packing state. Review Paper Scot. Inst. Agric. Eng. Biennial Rep, 58-86.

Soane, B.D. and Henshall, J.K., 1979. Spatial resolution and calibration characteristics of two narrow probe gamma ray transmission systems for the measurement of soil bulk density in-situ. J. Soil Sci., 30:517-528.

Stengel, P., 1979. Utilisation de l'analyse des systèmes de porosité pour la caractérisation de l'état physique du sol in-situ. Ann. agron. 30(1), 27-49.

Stengel, P. ; Bertuzzi, P. ; Gaudu, J.C., 1985. La double sonde gamma LPC-INRA. Précision, Utilisation agronomique. Bull. de liaison des Ponts et Chaussées.

Air permeability and gas diffusion measurements to quantify soil compaction

B.C.Ball

Scottish Institute of Agricultural Engineering, Bush Estate, Penicuik, Midlothian, UK

ABSTRACT

Air permeability and diffusion of Krypton-85 gas were measured in soil cores of differing bulk densities and degrees of saturation. Relative diffusivities reveal directly the conditions under which soil aeration becomes inadequate for arable crops. The spatial variability of air or water permeability may be estimated from air or water-filled porosities by using the relationship between air permeabilities and air-filled porosities. Relative diffusivity and air permeability may be used separately to calculate empirical continuity indices. Combination of the two measurements with air-filled porosities allows estimation of the radius, the variation of radius and the tortuosity of the continuous paths of soil pores. Measurements of samples taken at intervals after compaction show recovery from compaction which is not necessarily indicated by measurements of soil porosity or bulk density.

INTRODUCTION

Measurements of air permeability, relative diffusivity and air-filled porosity made on the same sample at a succession of matric potentials allow calculation of much information on the pore structure and its variation with moisture content. This is because the sample is a core of undisturbed soil of known dimensions and because permeability depends on the fourth power of pore radius whereas diffusion depends on the second power of pore radius or cross-sectional area. These methods thus allow the assessment of the reduction in pore size and continuity which usually result from compaction.

METHODS

The cores of soil are 73 mm in diameter and 50 mm long. The cores remained inside their stainless steel sampling cylinders during measurements. The relative diffusivity of Krypton-85 in air, and air permeability, were measured in each sample at several water potentials (Ball et al., 1981). Relative diffusivity is measured using the change in concentration of Krypton-85 during diffusion between gas chambers

15

attached to both end faces of the sample. Concentration is measured from regular counting of β radiation at a plastic scintillator-photo-multiplier tube assembly. Air permeability is measured by applying a pressure difference of <1.3 kPa across the sample and measuring the resultant steady-state laminar flow using a digital soap-film meter or rotameter. Dry bulk densities were calculated from the sample weight and volume.

SOILS AND TREATMENTS

Samples were taken from ploughed and direct drilled treatments in two field experiments, A and B, on Winton series (loam to sandy clay loam), a surface water gley (Ragg and Futty, 1967). The Proctor maximum dry bulk density is 1.38 Mg/m^3, the Proctor optimum water content is 21% w/w and the plastic limit is 31% w/w. With annual rainfall of 866 mm, the soil readily overcompacts under most arable crops if vehicle traffic is not controlled (Campbell, Dickson and Ball, 1982). Experiment A has treatments of conventional mouldboard ploughing to 250 mm and triple-disc direct drilling and receives normal random traffic only for seedbed preparation and harvest. Traffic at other periods is restricted to tramlines. Experiment B has treatments of chisel ploughing to 250 mm and triple-disc direct drilling. The direct drilling treatment was first applied in September 1981. All vehicle traffic is restricted to tramlines except for a seedbed traffic treatment applied each year before sowing. Traffic treatments reported here were zero, one pass of a standard tractor, applied uniformly across the entire plot width with recommended tyre inflation pressures and a similar single pass with tyre pressures reduced to half the minimum recommended values. This increases tyre-soil contact area and is intended to reduce the compactive effort of the tractor. The three treatments are referred to as Zero, 1 normal and 1 reduced respectively in the text. The two experiments and the experimental results are described more fully in (Ball and O'Sullivan, 1986) and (Campbell, Dickson and Ball, 1982).

In soil from experiment A, air permeability only was measured at -1, -6, -10, -19 and -109 kPa water potentials. In soil from experiment B, air permeability and relative diffusivity were measured at -6 kPa water potential. This corresponds to the air content at field capacity i.e. air capacity.

16

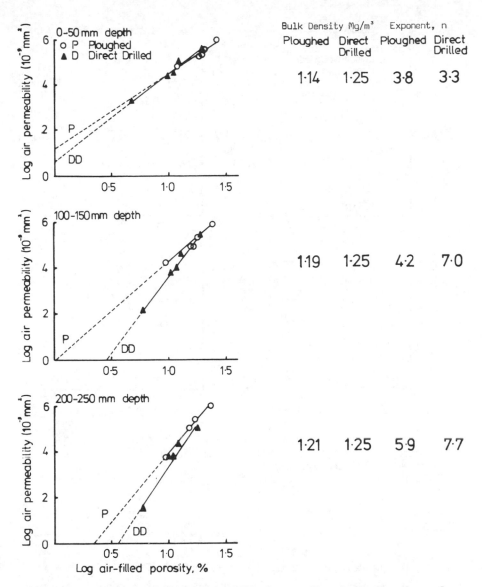

| Bulk Density Mg/m³ | | Exponent, n | |
Ploughed	Direct Drilled	Ploughed	Direct Drilled
1·14	1·25	3·8	3·3
1·19	1·25	4·2	7·0
1·21	1·25	5·9	7·7

Fig. 1. Air permeability, air-filled porosity, bulk density and exponent value (explained subsequently in the text) for soil from Experiment A at three depths. For the data points, the water potentials of equilibration are, from left to right, −1, −6, −10, −19 and −109 kPa.

RESULTS

   Arithmetic means over sampling positions are expressed as replicated relative diffusivities, air-filled porosities and dry bulk densities and geometric means for air permeabilities.

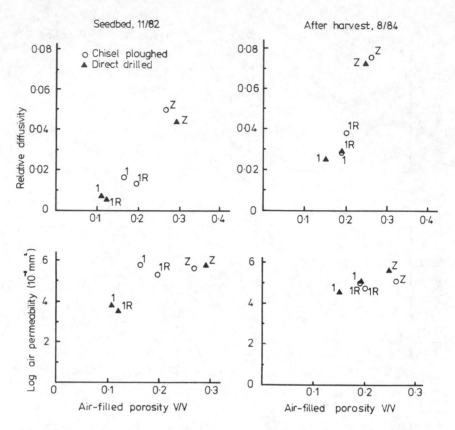

Fig. 2. Relative diffusivity, air permeability and air-filled porosity at air capacity (-6 kPa water potential) at 10-60 mm depth of soil from Experiment B.
Z = Zero pass, 1 = 1 pass, 1R = 1 pass reduced tyre pressure.

For experiment A, at a given water potential (Fig. 1), both air permeability and air-filled porosity were greater in ploughed than in direct drilled soil. For experiment B, air permeabilities, relative diffusivities and air-filled porosities (Fig. 2) were significantly greater at -6 kPa water potential in the zero than in the single pass treatments as might be expected from their lower bulk densities (Table 1). The cultivation and tyre inflation pressure treatments had only minor effects on any property.

18

Table 1. Bulk densities in the seedbeds and in the topsoil after
harvest in core samples from experiment B, Mg/m³ at 10-60 mm depth.

		Zero traffic	1 pass, reduced pressure	1 pass, normal pressure	s.e.
Seedbed	Direct drilled	1.02	1.37	1.32	0.04
(2 Nov. 1982)	Chisel ploughed	1.09	1.13	1.16	0.04
After harvest	Direct drilled	1.16	1.20	1.26	0.04
(30 Aug. 1984)	Chisel ploughed	1.12	1.18	1.22	0.04

DISCUSSION

Relative diffusivity provides a more relevant measure of aeration
status (Greenwood, 1975) than air-permeability or air-filled porosity
since it provides a measure of the rate of exchange of gas in the
structural pore space.   Stepniewski (1981) suggested that the lower
limit of soil aeration for plants lay between relative diffusivities
of 0.005 and 0.01.   The values after direct drilling in the seedbed in
November 1982 in experiment B (Fig. 2) are smaller than this range.
Separate measurements immediately below this depth at 60-110 mm reveal
that relative diffusivity in all the traffic treatments was low,
~0.0003 at air-filled porosities of ~0.08 v/v.   From Fig. 2 and other
data not given here these limiting relative diffusivities correspond to
air permeabilities of 10^{-5} and 10^{-6} mm².

The assessment of aeration status in soil from averages of air
permeabilities and relative diffusivities is restricted by the great
variability between measurements made on replicate samples.   For example
in the seedbed samples of experiment B, taken in November 1982,
permeability at -1 kPa water potential ranged from zero to 7×10^{-4} mm².
Half of the traffic treatment samples were impermeable to gas flow and
diffusion at -1 kPa water potential.   The air-filled porosities in such
samples, which can be considered non-continuous, ranged from 1.8 to
11.7% with an average of 5.4%.   Non-continuous porosity can thus be
estimated from these air-filled porosities in samples where gas movement
is zero at high water potentials.

When assessing the rootability and drainage status of a soil, the
potential of -6 kPa is an appropriate choice since the minimum radius of

pore drained is 25 μm and most root and water movement occur in pores of this size or greater. At -6 kPa, in experiment B, the greater relative diffusivities and air permeabilities (Fig. 2) at similar air-filled porosities at the second sampling in comparison to the first indicate that some structural improvement has occurred during that period, particularly in the direct drilled soil. Such improvement probably results from weathering and biological activity and is accompanied by a small decrease in bulk density (Table 1). Structural improvement can be quantified from the change in ratio (Table 2) of relative diffusivity to air-filled porosity (Ball, 1981b). This ratio is a measure of the ability of the pores to conduct gases and increases as the pore system becomes more continuous and/or less tortuous. Tortuosity in a soil core is the ratio of the actual distance through the pore system between the core faces to the shortest distance between faces.

$D/D_o \varepsilon_A$ increases between samplings, particularly after direct drilling and in the traffic treatments. Similar pore continuity indices were also proposed by Groenevelt, Kay and Grant (1984) as the quotient of air permeability and air filled porosity with further division by air-filled porosity to correct for any uniform increase in the radius of conducting pores with air-filled porosity. The two quotients calculated for experiment B (Table 2) differ between treatments and contrast with the diffusivity continuities in that pore radius has a large effect, particularly on K_A/ε_A, as well as continuity and tortuosity.

In order to get more details of pore size and number the continuous paths of soil pores can be treated as tubes and the radius and tortuosity of these tubes estimated. Tortuosity is the ratio of the total length of the tube to the length of the sample. Such tube modelling combines the diffusivities, permeabilities and air-filled porosities.

If the tubes are treated as single radius, then number, tortuosity and radius, can be calculated (Ball, 1981a). For the seedbeds of experiment B (Fig. 3), tortuosity tended to be greater in the direct drilled than in the ploughed soil and decreased between samplings.

Since tortuosity does not vary markedly with experimental treatments at a given potential, it can be assumed to be fixed, here at twice the sample length. This allows the application of a two-radius model of tubes which provides a measure of constriction, i.e. the ratio of the

Table 2. Pore continuities in seedbeds and in the topsoil after harvest at air capacity (-6 kPa matric potential) at 10-60 mm depth in Experiment B

			Zero traffic	1 pass, reduced pressure	1 pass, normal pressure
$D/D_0 \varepsilon_A$	Seedbed (2 Nov. 1982)	Direct drilled	0.147	0.043	0.065
		Chisel ploughed	0.182	0.066	0.101
	After harvest (30 Aug. 1984)	Direct drilled	0.297	0.157	0.165
		Chisel ploughed	0.292	0.191	0.154
$K/\varepsilon_A \times 10^{-6}$	Seedbed (2 Nov. 1982)	Direct drilled	22.5	0.29	0.58
		Chisel ploughed	14.2	10.8	36.1
	After harvest (30 Aug. 1984)	Direct drilled	18.6	5.7	2.6
		Chisel ploughed	5.5	2.9	6.4
$K/\varepsilon_A^2 \times 10^{-5}$	Seedbed (2 Nov. 1982)	Direct drilled	7.7	0.24	0.54
		Chisel ploughed	5.3	5.5	22.2
	After harvest (30 Aug. 1984)	Direct drilled	7.6	3.0	1.7
		Chisel ploughed	2.1	1.4	3.4

radii of interconnecting pores in a continuous path. This model is under development currently. The length of tube of each radius is in proportion to the tube radius. Constriction (the ratio of the two radii) and the larger tube radius (Fig. 3) decrease between sampling dates. The single tube radius model gives radii intermediate between those of the two radius model. The relatively large radii predicted by both models for the chisel ploughed 1 pass seedbed explain the high value of the continuity index K_A/ε_A^2 for these treatments. Such large pore radii ensure good drainage but are not necessarily suitable for providing good aeration and rootability since the surface area for gas exchange of large pores is relatively small (Greenwood, 1968) and such pores may not provide adequate support for roots, particularly where radius varies widely along the continuous pore path.

Data from experiment A fit the generalised Kozeny-Carman equation (Ahuja et al., 1984) which, if written for air permeability and air-filled porosity takes the form:

21

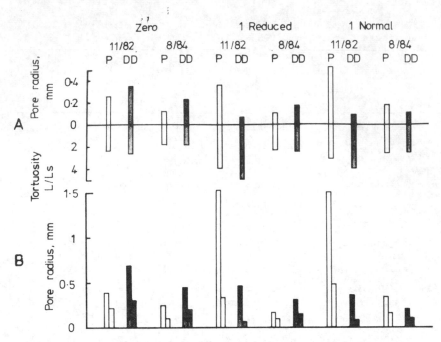

Fig. 3. Radii and tortuosities of tubes from a single radius model (A) and from a two radii model (B) applied to soil from Experiment B from 10 to 60 mm depth. In model B, tortuosity is assumed to be 2 times sample length except for zero pass 8/84 where it was 1.4 times sample length.

$$K_A = B\varepsilon_A^n$$

where B and n are empirical constants. Values for n (Fig. 1) ranged from 3.3 near the surface to 7.7 immediately below the plough layer. This relationship allows estimation of the spatial variability of air permeability from measurements of air-filled porosity (Ahuja et al., 1984). The spatial variability of hydraulic conductivity may be inferred from that of air permeability if, as is frequently observed, the two correspond closely.

CONCLUSIONS

1. Relative diffusivities may be used to identify the water potential and bulk density at which aeration may limit plant growth.

2. The ratio of relative diffusivity of air permeability to air-filled porosity reveals differences in pore continuity, constriction,

tortuosity and size which may be quantified empirically by
modelling the continuous paths of pores as tubes.

3.   Compacted soil tended to recover and loose soil to compact during
weathering and biological action in two seasons of direct drilling
such that the pore structure, as indicated by the results of
tube modelling, became similar in the two treatments.

4.   The Kozeny constant can be estimated from measurements of air permeability at different water potentials.   Thus the spatial variability of air or water permeability may be estimated from air- or
water-filled porosities.

REFERENCES

Ahuja, L.R., Naney, J.W., Green, R.E. and Nielsen, D.R. 1984. Macro-
porosity to characterise spatial variability of hydraulic conductivity and effects of land management.   Soil Sci. Soc. Amer. J.
48(4): 699-702
Ball, B.C. 1981a. Modelling of soil pores as tubes using gas
permeabilities, gas diffusivities and water release. J. Soil Sci.
32: 465-481
Ball, B.C. 1981b. Pore characteristics of soils from two cultivation
experiments as shown by gas diffusivities and permeabilities and
air-filled porosities. J. Soil Sci. 32: 483-498
Ball, B.C., Harris, W. and Burford, J.R. 1981. A laboratory method to
measure gas diffusion in soil and other porous materials. J. Soil
Sci. 32: 323-333
Ball, B.C. and O'Sullivan, M.F. 1986. Cultivation and nitrogen requirements for drilled and broadcast winter barley on a surface water
gley. Soil Till. Res. 6 (in preparation)
Campbell, D.J., Dickson, J.W. and Ball, B.C. 1982. Controlled traffic
on a sandy clay loam under winter barley in Scotland. Proc. 9th
Conf. Int. Soil Till. Res. Organ., Osijek, Yugoslavia, 1982
pp 189-194
Greenwood, D.J. 1968. Effect of oxygen distribution in the soil on
plant growth. In "Root Growth", ed. W.J. Whittington. Proc. 15th
Easter School of Agric. Science, Univ. Nottingham, Butterworths,
pp 202-223
Greenwood, D.J. 1975. Measurement of soil aeration. In "Soil Physical
Conditions and Crop Production" MAFF Tech. Bull. 29, HMSO, London
Groenevelt, P.H., Kay, B.D. and Grant, C.D. 1984. Physical assessment
of a soil with respect to rooting potential. Geoderma 34: 101-104
Ragg, J.M. and Futty, D.W. 1967. The soils of the country round
Haddington and Eyemouth. HMSO, Edinburgh
Stepniewski, W. 1981. Oxygen diffusion and strength as related to soil
compaction. II Oxygen diffusion coefficient. Polish J. Soil
Sci., 14: 3-13.

The importance of pore volume and pore geometry to soil aeration

H.-G.Frede

Institut für Bodenkunde, Universität Göttingen, FR Germany

ABSTRACT

There have been many attempts to describe the aeration status of soils only considering the aerated soil volume. A method is introduced, which permits consideration of both aerated soil volume and the structure of pores. The continuity of a single pore class is called specific pore continuity. Examples of the importance of pore structure for aeration are presented.

INTRODUCTION

Increasing mechanisation frequently leads to soil compaction and oxygen deficiency. Normally, the measurement of air-filled porosity is used in an attempt to describe soil aeration. However, this procedure does not take pore structure into consideration.

In the present study, an attempt is made to show the importance of pore structure for the aeration of soils.

THEORY

The aeration of soils is described by the diffusion equation:

$$1. \quad q = -D_s \cdot \frac{dc}{dx}$$

where

$$2. \quad D_S = D_0 \cdot \frac{1}{\tau} \cdot E_L$$

q = gas flux $\qquad\qquad$ $(cm \cdot t^{-1})$

D_S = diffusion coefficient $\qquad$ $(cm^2 \cdot t^{-1})$

dc = concentration gradient

dx = distance $\qquad\qquad$ (cm)

$\frac{1}{\tau}$ = pore continuity

E_L = porosity $\qquad\qquad$ $(cm^3 \cdot cm^{-3})$

D_0 = diffusion coefficient in air $\quad$ $(cm^2 \cdot t^{-1})$

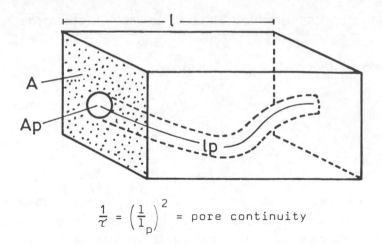

$$\frac{1}{\tau} = \left(\frac{1}{1_p}\right)^2 = \text{pore continuity}$$

Fig. 1 Tortuose pore

Equation 2 shows that the diffusion coefficient is – contrary to water flux – independent of pore diameter. In the following study however gas diffusion is considered as being dependent upon pore diameter. The continuity is called "specific pore continuity". There are two reasons for this procedure:

1. Gas diffusion and pore continuity must be considered in relation to pore diameter, because it is the soil water after all that opens or closes pores for aeration. The velocity of drainage is also dependent on pore diameter.

2. Pore continuities for different pore classes give information about pore geometry. They provide an excellent quantitative indicator of pore structure, giving information which otherwise becomes lost in the relationship between D_S/D_O and E_L, because this relation averages all continuities over the complete range of pores.

The specific pore continuity is calculated by:

3. $\quad \dfrac{1}{\tau}_{\text{spec.}} = \Delta D_S \;/\; \Delta E_L \cdot \dfrac{1}{D_O}$

METHODS

Undisturbed soil samples were taken in 100 ml cylinders, saturated under vacuum and placed on a ceramic plate extractor. After extraction, the sample was put into a diffusion chamber (Fig. 2). The air in the chamber was exchanged with N_2. Gas exchange now only was possible by diffusion through the soil sample. The increase of O_2 in the chamber was determined by gas chromatography.

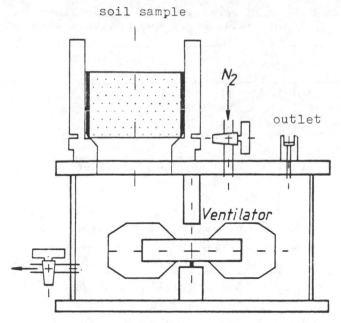

Fig. 2   Gas diffusion chamber

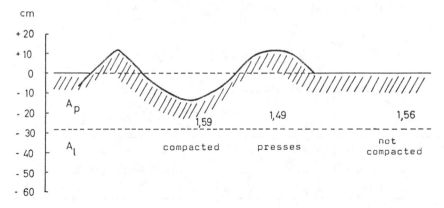

Fig. 3   Bulk densities under and beside a wheel track

RESULTS

In the following, an example is given to show the importance of the specific pore continuity for quantifying gas diffusion conditions in soils.

Figure 3 shows a cross section through a wheel track on a loess soil. Bulk density, total porosity, pore size distribution, apparent diffusion coefficients at different pF-steps and specific pore continuities were measured for this wheel track.

27

TABLE 1   Bulk density, total pore volume and pore diameter under and beside a wheel track

	bulk density g/ccm	total Vol.%	pore diameter					
			>300	300-50	50-10	10-5	5-3	< 3 µm
not compacted	1,56	41,1 b	4,8a	0,6a	2,4a	2,6a	2,5a	28,2 b
compacted	1,59	39,9a	3,9a	3,4 b	2,0a	1,8a	2,1a	26,7a
pressed	1,49	43,8 c	6,6 b	3,8 b	3,0a	2,2a	2,4a	25,8a

Means followed by the same letter are not significantly different at the 0.05 probability level.

TABLE 2   Diffusion coefficients at different pF-status under and beside a wheel track

	pF 1	pF 1,8	pF 2,5	pF 2,8	pF 3,0
not compacted	0,0042 b	0,0044 b	0,0047ab	0,0056a	0,0080 b
compacted	0,0022a	0,0028a	0,0039a	0,0056a	0,0062a
pressed	0,0028a	0,0030 b	0,0050 b	0,0063a	0,0075 b

Means followed by the same letter are not significantly different at the 0.05 probability level.

TABLE 3   Specific pore continuities under and beside a wheel track

	pore diameter				
	>300	300-50	50-10	10-5	5-3µm
not compacted	0,43 b	0,18 b	0,06a	0,17a	0,47 b
compacted	0,28a	0,09a	0,27 b	0,46 bc	0,15a
presses	0,21a	0,03a	0,33 b	0,29 b	0,25a

Means followed by the same letter are not significantly different at the 0.05 probability level.

Bulk density, total pore volume and macro-pores appear to indicate a good aeration beside the compacted wheel track (Table 1). This supposition is not confirmed by the measurement of gas diffusion, which indicates that the pressed part of soil beside the track is not better aerated than the compacted part (Table 2). The reason is given in Table 3, which shows the specific pore continuities. It shows that the continuity mainly of the big

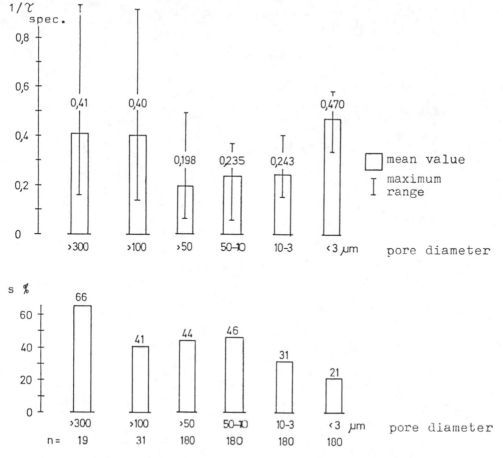

Fig. 4   Specific pore continuities of different Loess soils
          (o - 20 cm depth), mean values, maximum ranges and
          standard deviations

pores is reduced. Probably, these pores were sheared during pressing.

The specific pore continuity in different pore classes was studied in a
great number of samples of loess soils respectively. Figure 4 shows the
mean values, the maximum ranges and the standard deviations for all
classes. Macropores with a diameter greater than 100 µm have a high
continuity, but with a broad range of variation. In this class of pores,
which is mainly dependent on soil structure and independent of texture,
mechanical stress reduces continuity. On the other hand, biological
activity results in a high continuity. Pores of small diameter also have a
high continuity, probably because they only have a small path length of
capillary water inhibiting gas diffusion.

Micromorphometric and micromorphological investigations on the effect of compaction by pressures and deformations resulting from tillage and wheel traffic

M.Pagliai

Istituto per la Chimica del Terreno, C.N.R., Pisa, Italy

ABSTRACT

Porosity, pore shape and pore size distribution were measured on thin sections prepared from undisturbed soil samples by means of electro-optical image-analysis. Soil samples were taken from the topsoil compacted by the tractor wheels and in uncompacted topsoil, and just below the tilled layer where the ploughpan or ploughsole was developed.

In the compacted topsoil the porosity strongly decreased, particularly the irregular pores while the elongated ones strongly reduced their size and modified their orientation pattern. The microscopic observation revealed that these elongated pores were thin fissures parallel to the soil surface, thus giving rise to a platy structure, and very often they lost their continuity in a vertical sense. After one year there were no visible differences between the compacted and uncompacted topsoils. The ploughpans were very thick (in many cases more than 10 cm) but it was the first cm that was more strongly compacted and contained few, if any, pores.

INTRODUCTION

The first aim of tillage is the seed bed preparation and to create the best soil conditions for crop production, but in long-term intensive arable lands, it also produces damage to soil structure such as soil compaction. Soil compaction is caused by a combination of natural forces, which generally act internally, and by man-made forces related to the consequences of soil management practices. The latter forces are mainly those related to vehicular wheel traffic and tillage implements and have a much greater compactive effect than such natural forces as raindrop impact, soil swelling and shrinking, and root enlargement.

The aim of this experiment was to study the modifications of soil porosity and structure induced by tractor wheels in a clay loam soil and for how long such modifications could be identified visually. In addition, the ploughpan or ploughsole, formed by pressure and shear forces applied to the bottom of the plough layer in the same soil, was investigated.

MATERIALS AND METHODS

Undisturbed soil samples were taken from the Ap horizon of a clay loam soil. The soil was compacted by the normal traffic wheels of the tractor

31

four months after ploughing and the soil was left fallow for one year to study the differences between the compacted and the adjacent uncompacted topsoil during this year. Six topsoil samples were taken from each compacted and uncompacted soil just after the compaction, four, eight and twelve months after. Soil samples were also taken just below the topsoil (7-14 cm) and at depths from 25 to 45 cm for the ploughpan study. The samples were acetone-dried and made into 6 x 7 cm vertically oriented thin sections (Jongerius and Heintzberger, 1975). Photographs of thin sections (Ismail, 1975; Pagliai et al., 1984) were analysed by the image-analysing computer, Quantimet 720.

Measurements made on each sample included total porosity and the division of pores into three shape groups, i.e. rounded, irregular and elongated pores, and pores of each shape group were further subdivided into size classes according to either the equivalent pore diameter for rounded and irregular pores or the width for elongated pores. More detailed information on these porosity measurements are previously reported (Pagliai et al., 1983, 1984). Thin sections were also examined by a Zeiss "R POL" microscope at 40 magnification for micromorphological observations.

RESULTS AND DISCUSSION

Total porosity, expressed as a percentage of total area of thin section occupied by pores, was significantly higher in samples of uncompacted soil than in those of compacted soil just after the compaction by tractor wheels (Fig. 1).

The pore size distribution showed a strong reduction in the size of pores in the compacted soil and also showed modifications in the pore shape. The higher area proportion of pores in the compacted soil was represented by elongated pores; the irregular pores strongly decreased with respect to the uncompacted soil. The microscopic observations showed that the thin elongated planar pores traversed the soil material in a fairly regular pattern parallel to the soil surface producing a rather compact platy structure (Fig. 2). In large areas the continuity of these pores in a vertical sense was practically absent. The small rounded pores were distributed within the platy structural units. On the contrary in the uncompacted topsoil the pores had a more random arrangement and a more uniform distribution, thus producing a complex microstructure: part moderately developed subangular blocky and part vughy, i.e. the peds were not separated and the mass was broken up by scattered irregular pores ("vughs", Brewer, 1964) and occasionally by channels and chambers.

32

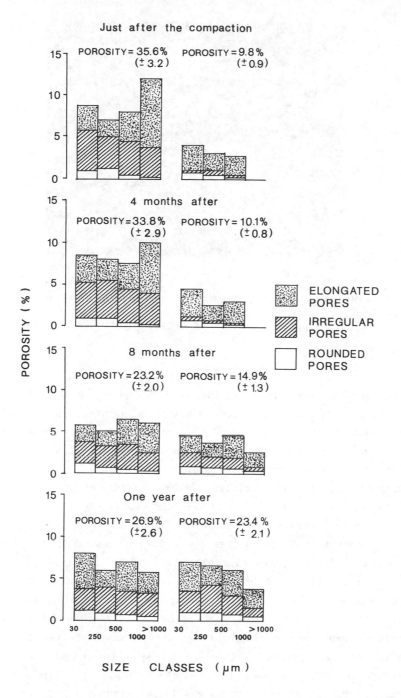

Fig. 1 – Porosity and pore size distribution according to either the equivalent pore diameter for regular and irregular pores or the width for elongated pores; means of six replications.

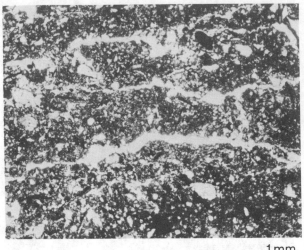

1mm

Fig. 2 - Microphotograph of vertically-oriented thin section of soil
samples from the surface layer (0-7 cm) of the compacted soil just after
the compaction. A platy structure is very evident. Plain light.

Four months from the beginning of the experiment the porosity, pore
shape and pore size distribution did not show any significative differences
with respect to the previous sampling both in the uncompacted and compacted
soil. Also the microscopic observation did not reveal variations in the
microstructure of the two sets of samples.

On the contrary the results obtained after eight months from the
beginning showed a modification in the soil conditions. In the uncompacted
soil the total porosity significantly decreased with respect to the
previous sampling and this decrease was due to the irregular pores and to
the larger elongated pores (larger than 1 mm). Since this sampling was
made in March, this could be explained as the result of physical stresses
caused by heavy winter rainfall that increased soil compactness, reducing
principally the large pores. A different trend was observed in the
compacted topsoil where the total porosity significantly increased with
respect to the previous sampling, even though the porosity was
significantly less than in the uncompacted soil. This increase was
principally due to irregular pores and larger elongated pores (>1 mm) which
were absent in the previous sampling. In this case the wetting and drying
cycles could have contributed to the formation of these pores and

34

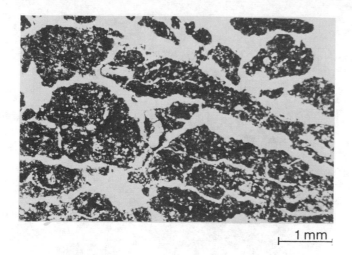

⊢ 1 mm ⊣

Fig. 3 - Microphotograph of vertically-oriented thin section of soil
samples from the surface layer (0-7 cm) of the compacted soil after one
year from the compaction. A vughy to subangular blocky structure is
visible. Such a situation is very common also in the uncompacted soil.
Plain light.

consequently to the beginning of the regeneration of soil structure. The
microscopic observations showed, in fact, a less compact soil structure.

After one year the total porosity did not show any significative
differences between the uncompacted and the compacted soil. Therefore in
the compacted topsoil the porosity strongly increased with respect to the
previous sampling. Also in the uncompacted topsoil the porosity increased
but not significantly with respect to the sampling of March (four months
before).

As was the case with total porosity the pore size distribution of the
uncompacted topsoil was quite similar to that of the uncompacted topsoil
apart from a lower proportion of pores larger than 1 mm and particularly a
lower proportion of large irregular pores in the compacted topsoil.

Also the microscopic observations did not reveal measurable differences
in both sets of soil samples. The platy structure of the compacted topsoil
just after the compaction was transformed into a vughy to subangular blocky
structure as in the uncompacted soil and pores had a more random
arrangement and a more uniform distribution (Fig. 3).

As a concluding remark these data suggest that in this kind of soil
(clay loam soil) it would be better to reduce, as much as is practically

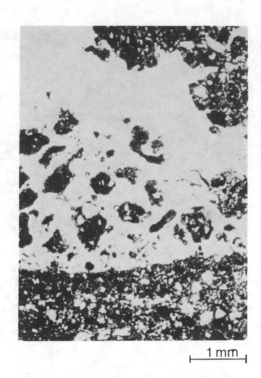

⊢ 1 mm ⊣

Fig. 4 - Microphotograph of vertically-oriented thin section of soil sample from the subsoil (28-35 cm). The limit of cultivation and the upper part of the ploughsole are visible. Plain light.

possible, the passage of heavy machinery which may cause a strong soil compaction with a consequent slowing down of the soil structure regeneration.

The thin sections prepared from soil samples, collected at depths from 25 to 45 cm four months after the last ploughing, showed that a ploughpan or ploughsoil was well developed and was formed by pressure and shear forces applied to the bottom of the plough layer. Fig. 4 shows the clear limit of the cultivation which occur to about 30 cm depth. The micromorphometric analysis showed that the porosity just above the ploughsole was still high and did not show significative differences with respect to the sample taken at 7-14 cm depth as reported in Fig. 5. This figure also shows that the effect of the wheel compaction reached a depth of about 10 cm. Also in the surface layer (0-5 cm) of the uncompacted soil the porosity was less than in the layer below 5 cm. This could be ascribed to the natural compactness of the soil due to, for example, the effect of raindrop impact.

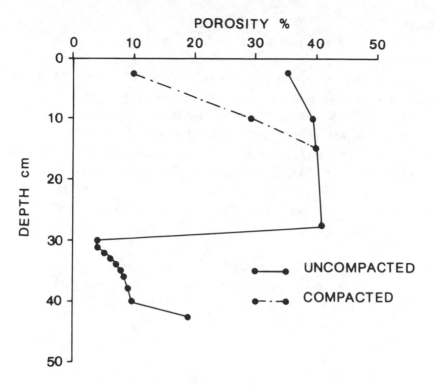

Fig. 5 - Porosity as percentage of total area of
thin section at selected depths.

In the ploughsole the porosity strongly decreased to a value of 4.0% in
the first cm then slightly increased but the micromorphometric analysis
revealed that the ploughsole can reach a thickness even of 10 cm. The
microscopic observations revealed that in the first cm the few pores
present were very thin elongated pores parallel to the surface of the
ploughsole. Some irregular pores were present in the lower part of the
ploughsole.

The presence of such a compacted layer at the lower limit of
cultivation in the profile may strongly reduce the water drainage and
hamper root development at depth.

Acknowledgements

The author wishes to thank Mr. M. La Marca, Mr. G. Lucamante and Mr. G.
Navarra for technical assistance.

37

REFERENCES

Brewer, R. 1964. *Fabric and Mineral Analysis of Soils*. John Wiley, New York, 470 pp.

Ismail, S.N.A. 1975. Micromorphometric soil porosity characterization by means of electro-optical image analysis (Quantimet 720). Soil Survey Paper No. 9, Netherland Soil Survey Institute, Wageningen, 140 pp.

Jongerius, A. and Heintzberger, G. 1975. Methods in soil micromorphology. A technique for the preparation of large thin sections. Soil Survey Paper No. 10, Netherlands Soil Survey Institute, Wageningen, 48 pp.

Pagliai, M., La Marca, M. and Lucamante, G. 1983. Micromorphometric and micromorphological investigations of a clay loam soil in viticulture under zero and conventional tillage. *Journal of Soil Science* 34, 391-403.

Pagliai, M., La Marca, M., Lucamante, G. and Genovese, L. 1984. Effects of zero and conventional tillage on the length and irregularity of elongated pores in a clay loam soil under viticulture. *Soil and Tillage Research* 4, 433-444.

Observation morphologique de l'état structural et mise en évidence d'effets de compactage des horizons travaillés

H.Manichon

Institut National Agronomique Paris-Grignon, France

ABSTRACT

Compaction and fragmentation actions of agricultural machines generally don't affect the whole volume of Ap horizon. The result is a greater spatial variability of soil structure within the ploughed layer than in untilled horizons, leading to some difficulties to characterize, at field, soil physical conditions.

The method presented here consists of : (i) drawing up a vertical and lateral partition of soil profile, taking into account the depth of action of tillage tools (which definite several anthropic horizons) and the positions where tractor wheels have passed or not ; (ii) describing one or several morphological units in each zone then being defined.

Soil tilth in a morphological unit is defined by the combination of 2 criteria : (i) visible porosity of clods : "Δ" is compact, containing only textural pores ; "Γ" contains structural pores, aggregates are visible in clods ; "Φ" derives from "Δ" but contains cracks ; (ii) the way the clods are brought together : 4 types can be defined ("M" when massive, "SD" when clods are very closely packed together, "SF" when they are adherent but can be individualized, "F" when they are not adherent – diameter of clods is noted in F and SF).

A soil tilth map can then been drawn up in each anthropic horizon : it is a usefull guide for physical measurements in such heterogeneous horizons, and for soil-roots interactions studies. This map can be synthetized : (i) evaluation of "Δ" frequency, (ii) 3 main types of horizons are distinguished according to domination of F and SF – with small clods – ("O" type), M and SD ("C" type), M and F – with decimetrical clods separated by large voids – ("B" type).

This approach leads to an understanding of soil structure formation, as a result of climatic agents and agricultural practices interactions ; particularly, recent and cumulative compacting effects on soil can be distinguished. "Δ" frequency (which can be obtained also by densimetric screening) is a global index of crop rotation and machines effects on soil structure, in interaction with climate and texture.

INTRODUCTION

La couche supérieure du sol, dans les parcelles cultivées, subit des contraintes mécaniques variées, exercées par les machines et les agents naturels. Leurs effets sont contradictoires (fragmentations et compactages) et inégalement répartis (création d'horizons anthropiques d'épaisseur variable, roues des engins n'affectant qu'une partie du volume du sol, effets des

39

agents climatiques surtout intenses au voisinage de la surface...). La variabilité spatiale de l'état physique que l'on constate dans le profil cultural (Hénin et al., 1969), grâce à la répétition de profils de densité (Soane et al., 1971) ou de résistance pénétrométrique (Billot, 1982), résulte de ces différents effets.

Mais les représentations cartographiques tirées de la répétition systématique de ces mesures sont d'une interprétation délicate, du fait de l'influence de variations de teneur en eau à l'échelle décimétrique (Papy, 1984) comme de l'impossibilité de faire correspondre, à une valeur donnée du paramètre mesuré, un état structural et une origine définis : l'emploi de méthodes morphologiques est nécessaire pour élucider l'origine de l'état du profil, et en tirer un diagnostic sur les effets du système de culture. L'expérience nous a montré (Manichon, 1982a) que l'adoption d'une attitude déterministe, débouchant sur la définition de critères et de leurs procédures d'évaluation, permettait de réduire les risques de subjectivité inhérents aux approches qualitatives. Nous présentons ici la méthode que nous avons conçue, et illustrons son emploi pour la mise en évidence d'effets de compactage d'origines diverses, dans les horizons travaillés.

UNE METHODE D'ETUDE DU PROFIL CULTURAL

De par ses origines, la variabilité spatiale de l'état structural n'est pas, essentiellement, de nature aléatoire. Dès lors, il est possible de procéder à une partition du volume de sol examiné in situ, sur la base de causes de variation connues a priori.

Partitions verticale et latérale du profil

Pour exposer les principes de ces partitions, nous prendrons l'exemple d'une parcelle observée après semis de la culture, et dont le travail du sol comportait un labour (Fig. 1). L'analyse des autres situations se déduit facilement de cet exemple.

Sur la paroi verticale d'une fosse, on distingue, outre les horizons pédologiques, plusieurs horizons anthropiques dans Ap. Les variations brutales de l'état structural dans le sens vertical, les traces des outils (lissages), sont les symptômes qui guident pour réaliser cette partition.

Mais chaque horizon est susceptible de présenter en son sein une certaine variabilité. Celle-ci est due :

. **au labour lui-même** : création de discontinuités par le découpage de bandes de terre, comportements mécaniques contrastés de celles-ci, liés à la préexistence d'états de compacité (ou d'états hydriques) différenciés,

. **aux roues des engins ayant circulé après le labour** (formation de zones compactes localisées, dans certaines conditions).

Ceci n'est nettement visible qu'en l'absence de fragmentation postérieure à ces actions. C'est le cas pour H5, par ailleurs moins modifié depuis les travaux par l'action des agents climatiques que les horizons supérieurs ; on accordera à cet horizon une attention particulière.

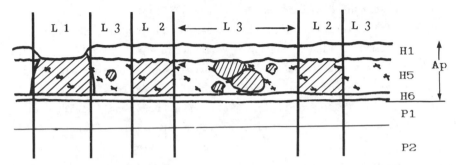

Fig. 1 Partitions verticale et latérale du profil cultural *(vertical and lateral partitions of soil profile)* : P1, P2... : Horizons pédologiques *(pedologic horizons)* ; H1, ... H6 : Horizons anthropiques (H5 : partie de la couche labourée non fragmentée par les travaux postérieurs) *(anthropic horizons within Ap ; H5 : part of ploughed layer which has not been fragmented by secondary cultivation)*. L1, L2, L3 : Positions latérales *(lateral positions)*.

A partir de ces considérations, 3 types de **positions latérales** peuvent être définies, qui correspondent à des étapes différentes des interventions culturales :

. **L1** : il s'agit des emplacements affectés par des roues, après les derniers travaux du sol, leurs traces sont visibles en surface ;

. **L2** : il s'agit des endroits où ont circulé les roues des engins utilisés entre le labour et la dernière façon d'ameublissement ; s'ils n'ont pas été repérés au moment des travaux, il est possible de déduire leur position de la connaissance précise des interventions culturales, mise en relation avec certains traits morphologiques de H5 (zones plus compactes, dont les localisations rendent plausible cette interprétation) ;

. **L3** : c'est un résidu, indemne des actions précédentes.

Les limites des positions latérales (tracées verticalement par simplification) recoupent celles des horizons : on obtient un découpage de la face observée en plusieurs **zones**. Cette opération ne définit que le cadre de la description : elle n'implique ni l'identité entre zones d'un même type de position, ni l'existence de différences entre positions.

L'unité morphologique et sa description

Certaines des zones ainsi définies apparaissent visuellement homogènes, d'autres non (Fig. 2) ; on peut alors y distinguer plusieurs **unités morpho-logiques**. Cette notion n'a de sens qu'en référence à un système de description permettant de positionner de façon non ambiguë les limites entre des volumes présentant des états structuraux différents.

Les classifications de la structure basées sur la dimension des éléments et la forme de leurs contours (travaux de Nikiforoff, 1945, plus récemment précisés dans leurs modes d'évaluation au champ par Hodgson, 1974), utilisées en Pédologie, visent surtout à révéler le résultat des processus naturels de structuration. Cette finalité rend malaisée et peu pertinente leur utilisation dans les horizons travaillés, constitués pour l'essentiel d'éléments structuraux (que nous appellerons "mottes" pour les distinguer des agrégats "naturels"), dont les caractères et l'agencement sont surtout liés à des actions anthropiques perturbatrices. Pour définir des critères d'évaluation révélant ces actions, il faut expliciter leurs rôles au cours des interventions culturales successives.

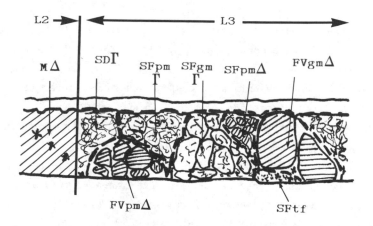

Fig. 2 Unités morphologiques en L2 et L3 de H5 (exemples)
(Morphological units in L2 and L3 within H5 - Examples)

Bien que l'on soit encore très loin de pouvoir modéliser efficacement le comportement mécanique de l'ensemble de la couche labourée sous le jeu combiné de compactages et de fragmentations, il est possible de tenir compte de comportements élémentaires. Prenons un état initial constitué d'agrégats de petite taille ; selon l'état hydrique de la terre et l'énergie appliquée, un compactage entraîne un réarrangement de ces constituants de la structure (diminution de l'espace poral entre eux) ou leur déformation, pouvant aller jusqu'à la coalescence et la disparition de la porosité structurale (Guérif, 1982) ; une fragmentation postérieure découpe le volume compacté en éléments d'autant plus gros que le compactage avait été marqué (Manichon, 1982a) : nous dirons que les actions anthropiques successives, selon leur intensité et la consistance de la terre, peuvent affecter **deux niveaux d'organisation structurale** (Manichon, 1982b) :

. **l'état interne des mottes** : leur cohésion et l'aspect de leurs faces de rupture, après fragmentation manuelle, permet de distinguer 2 états typiques, de valeur assez générale :

- l'état "Γ" correspond à une face rugueuse et à une porosité visible importante, due aux défauts de contact entre les agrégats constitutifs ; la morphologie de ceux-ci varie selon la texture (cet état paraît assez directement en relation avec les processus naturels de structuration) ;

- pour l'état "Δ", la cohésion est plus élevée, les faces de fragmentation sont lisses, d'aspect continu et de forme conchoïdale, sans porosité visible ; cela correspond à la coalescence des agrégats préexistants, sous l'action de contraintes mécaniques sévères ;

L'observation de l'état interne n'est, pratiquement, possible au champ que sur les mottes ayant une certaine dimension ($>$1 cm de diamètre environ).

. **Le mode d'assemblage des mottes** : on peut définir les modalités suivantes:

- lorsque plusieurs mottes de dimensions comparables coexistent dans un volume donné, on parle d'état **fragmentaire**, en distinguant les états "F" (mottes indivudalisées) et "SF" (mottes adhérentes) ; la dimension des mottes est notée, ainsi que les cavités les séparant (suffixe "V") ;

- une unité morphologique peut être constituée d'un seul élément de grande dimension, on parle d'état **massif** ("M") ;

- entre ces cas extrêmes existe celui où les mottes créées par les outils sont devenues **difficilement discernables**, sans qu'on puisse parler d'un état continu (état "SD").

Les états observés au champ résultent aussi des agents naturels interagissant avec la texture ; il faut donc examiner la manière dont les 2 niveaux d'organisation définis ci-dessus en sont affectés. Dans l'état actuel des connaissances, on peut admettre que :

. le plus souvent, l'état interne n'est pas modifié par les agents naturels, les cas d'exception concernant :

- la surface et les horizons les plus superficiels : en terres battantes, la désagrégation des éléments sous l'action des pluies entraîne un remplissage des vides structuraux (Boiffin, 1984) et l'apparition de structures massives d'état interne morphologiquement voisin de "Δ", mais reconnaissable par son aspect stratifié ; les horizons sous-jacents (et notamment H5) ne semblent pas subir cette évolution, au moins sur des volumes appréciables ; on a cependant constaté la création de l'état "Δ" à partir de terres limoneuses très affinées, après submersion par irrigation et dessiccation ;

- certaines terres argileuses : prise en masse lors de la dessiccation de petits agrégats (Monnier & Stengel, 1982), fissuration d'éléments compacts sous l'action du gel (ou de la dessiccation). On est conduit, pour ce dernier cas, à définir une nouvelle modalité : l'état "Φ" proche de "Δ", mais comprenant des amorces de fissures ; il n'apparaît nettement qu'à partir d'un certain taux d'argile (environ 20%).

Sous ces réserves, les voies de passage entre états internes, naturelles ou anthropiques, peuvent être inventoriées (Fig. 3) ; ce schéma suggère que selon le système de culture (intensité et fréquence des fragmentations et des compactages) et selon le milieu (interactions Climat-Texture), on n'aura pas les mêmes fréquences des différents états internes ;

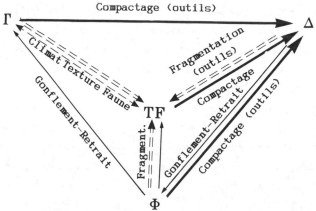

Fig. 3 Voies de passage entre états internes (Manichon, 1982a)
(*Relationships between typical internal status of clods*)

. le mode d'assemblage est plus contingent : si la dessiccation entraîne
l'apparition de symptômes identifiables (les fentes de retrait, on utilise
le suffixe "R" pour noter le passage de "M" à "MR", par exemple),
l'évolution sous l'action des pluies n'est pas distinguable de celle
obtenue par différentes intensités de compactage. Elle correspond à une
diminution progressive de la porosité structurale entre mottes (moins
rapide lorsque les mottes sont d'état interne "Δ") : passage de "F" à
"SF", "SD" ou "M". Ainsi, la séparation des effets des agents naturels de
ceux des actions culturales est plus ambiguë quand on considère le mode
d'assemblage plutôt que l'état interne.

La mise en oeuvre de cette grille de classification à 2 niveaux permet
de caractériser l'état structural de chaque unité morphologique, et d'en
préciser les limites avec les voisines (changement de modalité pour l'un des
niveaux au moins). Lorsqu'une certaine quantité de terre fine (éléments de
taille inférieure à 5 mm) est associée aux mottes, il est utile de rapporter
la surface occupée par les mottes (et donc par un état interne donné) à la
surface de l'unité morphologique : ceci fournit une évaluation du degré
d'affinement et permet une mise en correspondance avec les résultats des
tris densimétriques (cf. infra).

Synthèse des observations

Selon les contraintes qu'il a subies, un même matériau peut donc se
présenter sous différents états morphologiques ; ceux-ci coexistent dans le
même horizon, quand celui-ci n'a pas subi pour tout son volume la même
histoire de contraintes. La carte de l'état structural, établie selon la
procédure que nous avons présentée, est une représentation schématique de
cette variabilité. La synthèse de ces données, pour chaque position latérale
d'un horizon, se fait en évaluant la fréquence de chaque état structural par
sommation des surfaces des unités morphologiques concernées, rapportée à la
surface cartographiée. On en tire :
. **Les fréquences des états internes** ou, plus simplement, celle de l'état
"Δ". On peut aussi évaluer ce critère sur échantillons prélevés au champ,
par tri densimétrique et pesée : ceci est rendu possible par le fait que
la densité de ces éléments est peu variable pour une même terre (les
éléments "Φ", de densité très proche, ne peuvent en être séparés),
contrairement à celle des éléments "Γ". Il ne s'agit pas d'une simple
tendance : d'une part nous avons constaté, pour une terre limoneuse de

GRIGNON (Yvelines), une grande similitude entre la densité texturale (Fies, 1971) et la densité sèche de mottes " Δ " (1.65 et 1.66 respectivement) ; d'autre part, sur une série de 24 matériaux naturels contenant 12 à 30% d'argile, nous avons obtenu une régression (ρ_d = 0.013 A + 1.53 ; r = 0.81 (HS), avec ρ_d : masse volumique sèche des mottes " Δ " (moyenne d'une vingtaine de mesures pour chaque terre) ; A = teneur pondérale en argile), dont les coefficients sont très proches de ceux calculés par Fies et Stengel (1981) pour la densité texturale (0.015 et 1.45 respectivement). Ces faits tendent à montrer que la morphologie "Δ" correspond, pour ces matériaux, à l'absence d'un espace poral structural, et que l'observation visuelle permet de saisir la limite avec le domaine textural.

. **Les fréquences des différents modes d'assemblage.** Il est commode, pour faciliter la comparaison de plusieurs situations, de regrouper les modalités voisines : on définit alors un **troisième niveau d'organisation structurale**, comportant 3 types : "O", correspond à une dominance de F et SF (avec une certaine quantité de terre fine et sans grandes mottes décimétriques, ni cavités importantes) ; **en "B"**, dominent M et F (avec mottes décimétriques séparées par des cavités) ; "C" ne comporte pas de discontinuités structurales notables (M et SD dominent).

MISE EN EVIDENCE D'EFFETS DE COMPACTAGE DE DIVERSES ORIGINES

Deux catégories d'utilisation correspondent à l'observation de profils culturaux selon cette méthode :

. caractériser l'état structural, dans sa variabilité spatiale à l'échelle décimétrique à métrique, comme **"variable d'entrée"** pour des travaux au champ où cette donnée interfère sur les processus étudiés : disposition et fonctionnement des racines (Tardieu, 1984 ; Tardieu & Manichon, 1986a et b), comportement mécanique ou hydrique des couches travaillées (Papy, 1984) ; ceci complète, dans le domaine structural, l'analyse des systèmes de porosité (Stengel, 1979) ;

. caractériser un état comme **résultat d'une histoire** : c'est ce second aspect que nous allons illustrer ici. Notons qu'il ne s'agit pas de constater simplement un changement brutal sous l'action d'un outil (ce qui n'impliquerait que la comparaison de 2 états successifs), mais d'analyser le résultat d'actions multiples, récentes ou plus anciennes.

Principes d'interprétation

A partir de nombreuses analyses de profils culturaux et de techniques culturales (Manichon, 1982a ; Manichon & Sebillotte, 1973, 1975 ; Manichon & Bodet, 1976 ; Manichon et al., 1985), on peut tenter un essai de synthèse des effets des outils sur le sol, indispensable pour les interprétations :

. **Effets du labour** : l'humidité de la terre au moment du labour n'interfère **nettement sur le degré de fragmentation obtenu que si l'état initial n'était pas dégradé** : si l'état interne "Γ" domine, on obtient après labour à humidité moyenne un état "$0\,\Gamma$" (d'autant plus émietté que la vitesse d'avancement était rapide et la texture moins argileuse), et un état "$B\,\Gamma$", si cette opération a été effectuée en conditions très sèches (comportement rigide de la terre) ou très humides (comportement plastique) ; inversement, si "Δ" domine dans l'état initial, on obtient un état "$B\,\Delta$" après labour, quelle que soit l'humidité de la terre.

Il semble aussi que **l'énergie de compactage de la charrue soit trop faible pour entraîner la formation de l'état "Δ"**, au moins sur des volumes appréciables (à l'exception - peut-être - de disques utilisés en conditions plastiques) : **l'état interne "Δ" observé dans les positions L3 est ainsi, pour l'essentiel, hérité de l'état avant labour.** C'est sur cette base qu'on peut interpréter la relation stricte observée, en sols limoneux, entre le taux de l'état interne "Δ" et l'importance de l'état "B" (Manichon et al., 1985).

. **Effets des travaux post-labour** dans les positions L1 et L2 de H5, on note, au mieux, une conservation de l'état créé par le labour (en dehors des actions climatiques) ; le degré de compactage observé dépend de l'énergie appliquée et de l'humidité de la terre, mais les états "B" résistent mieux au compactage que les états "0" (Papy, 1984). Pour les compactages modérés, seul l'assemblage des mottes est modifié, les compactages sévères peuvent créer l'état interne "Δ".

On est alors conduit à considérer comme impossibles certaines combinaisons des états des diverses positions d'un même horizon (par exemple, si L3 est de type "0", L1 et L2 ne peuvent être de type "B", la proportion de "Δ" ne peut être plus faible en L1 et L2 qu'en L3) ; de tels cas peuvent révéler des partitions erronées, il faut alors agrandir la longueur de la face d'observation pour les contrôler ; ceci est particulièrement utile quand, l'état après labour étant composite (présence simultanée des types "0" et "B", en relation avec des compactages localisés préexistants), se trouvent

47

par endroit des coïncidences de position entre zones de type "B" et passages d'engins après labour.

Utilisant ces données, on peut comparer des situations dans lesquelles les énergies appliquées ou les états initiaux des parcelles étaient contrastés. On manque à l'heure actuelle de références suffisamment précises sur les seuils de comportement mécanique pour pouvoir interpréter tous les cas. Ces références sont particulièrement difficiles à établir, car elles doivent tenir compte à la fois des gradients hydriques (Guérif, 1984) et de la variabilité spatiale de l'état structural.

Distinction des effets de compactage actuels et hérités

Par simplification, nous supposerons ici que chaque position latérale n'est redevable que d'un seul type d'état structural. Les différentes combinaisons possibles et les interprétations qui en sont tirées sur les effets

Tab. 1 Les états des positions latérales et leur interprétation
(Soil structure in lateral positions and their interpretation)

État	POSITION L3 INTERPRETATION (Etat initial Eo et effet du labour)		État	POSITIONS L1 ET L2 INTERPRETATION (Effets des travaux post-labour)	
0 Γ	Eo non dégradé, fragmentation intense par le labour		0 Γ c Γ c Δ	Pas de dégradation Dégradation modérée Dégradation intense	
0 Δ	Eléments initiaux "Δ" de faible calibre (dégradation ancienne ou fissuration préalable)		0 Δ c Δ	Pas de dégradation Dégradation modérée ou intense	
B Γ	Eo non dégradé, fragmentation limitée due aux conditions de labour		B Γ c Γ c Δ	Pas de dégradation Dégradation modérée Dégradation intense	
B Δ	Eo dégradé		B Δ c Δ	Pas de dégradation Dégradation modérée ou intense	
c Γ	Eo non dégradé, absence de fragmentation (pas de labour récent)		c Γ c Δ	Pas de dégradation Dégradation intense	
c Δ	Eo dégradé, absence de fragmentation (pas de labour récent)		c Δ	Pas de dégradation, ou dégradation modérée, ou dégradation intense	

des interventions culturales antérieures à la date d'observation, sont présentées au Tab. 1.

On note :

. qu'un même état peut faire l'objet d'interprétations différentes, selon l'état des positions voisines et qu'ainsi une observation limitée à un volume trop restreint pourrait conduire à des interprétations erronées (pratiquement, il est nécessaire d'observer une longueur de 2 à 3 m) ;

. que les possibilités de diagnostic sur les compactages actuels sont conditionnées par l'intensité des compactages hérités, et inversement ;

. que, dans de nombreux cas, la comparaison des états structuraux des différentes positions latérales permet un diagnostic sur les effets des opérations culturales successives, c'est-à-dire sur les éléments de l'itinéraire technique (Sebillotte, 1978) de l'année en cours : **sans la caractérisation d'une variabilité spatiale à plusieurs échelles, seuls des effets globaux, difficilement interprétables, auraient pu être mis en évidence.**

Evaluation des effets cumulatifs liés aux systèmes de culture

C'est naturellement l'état interne de H5 en L3 qu'il faut considérer ici, cette donnée étant peu contingente de la date d'observation et non affectée par les dégradations post-labour de la campagne culturale en cours. Les résultats du Tab. 2 concernent plusieurs successions de cultures existant au centre d'Expérimentation de GRIGNON (Yvelines), les pourcentages de " Δ " (et " Φ ") ont été obtenus par tri densimétrique. On constate que ces valeurs varient du simple au double entre les cas 1, 2 et 3 d'une part, 4 d'autre part : la présence, tous les 2 ans, d'une culture de maïs dans ce dernier cas, par les compactages qu'elle entraîne au moment de la récolte, semble responsable de ce niveau élevé.

Tab. 2   Teneurs pondérales en " Δ " dans H5 (L3) pour différentes successions de cultures à GRIGNON (Yvelines) (" Δ " *frequency - densimetric screening - in H5 (L3) for different crop rotations*)

Successions de cultures	% "Δ"
1. Jachère travaillée	30
2. Luzerne-Blé-Colza-Blé	33
3. Colza-Blé-Maïs-Blé-Blé	34
4. Maïs-Blé (depuis 10 ans)	68
Succession 3 + Tassement avant labour	66
Positions L1 (toutes successions)	96

49

Cette interprétation est confortée par la valeur trouvée pour la succession 3, quand celle-ci a reçu avant le dernier labour un tassement homogène simulant l'effet d'une récolte tardive. Elle est confortée aussi par l'analyse du bilan de l'état interne " Δ " (effets des opérations de récolte exclus) ; dans ces parcelles, travaillées avec les mêmes outils de faible largeur, les emplacements L1 et L2 sont confondus, la répétition des compactages entraîne une transformation totale en " Δ ", quelle que soit la culture implantée (Tab. 2) ; en supposant que, dans ce milieu, seul l'horizon superficiel ameubli après labour est le siège de disparition de mottes " Δ " sous l'action des outils et du climat, on peut estimer à environ 30% la teneur en " Δ " de L3, après quelques années d'application de ces techniques. Cette valeur est très proche de celles trouvées pour les successions 1 à 3 ; l'accroissement constaté dans la succession 4 semble donc bien lié aux récoltes du maïs, les récoltes des autres cultures ne paraissant pas pouvoir entraîner des dégradations du sol.

CONCLUSION

Le dernier exemple cité montre, à la fois, le parti qu'on peut tirer de ces modes de caractérisation de l'état structural des horizons travaillés, pour une interprétation des effets des systèmes de culture, et le chemin qui reste à parcourir pour aboutir à la connaissance précise (et à la prévision) du déterminisme de l'évolution de l'état physique d'une parcelle cultivée.

En effet, le rôle des agents naturels étant connu par l'évaluation des propriétés liées à la constitution du matériau (Monnier & Stengel, 1982), il est nécessaire de prendre en compte de façon explicite le rôle de l'état structural vis-à-vis de ces actions (Boiffin, 1984) ; quant à l'influence des facteurs culturaux, si des effets globaux liés aux successions de cultures peuvent être énoncés, l'analyse fine de ceux des itinéraires techniques en est à ses débuts. Or ceux-ci, par les caractéristiques dimensionnelles des engins employés, leurs réglages (vitesse et profondeur) et leurs conditions d'emploi, sont responsables en grande partie des états observés. On a vu comment la méthode présentée dans cet article permettait analyses et diagnostics, à l'échelle stationnelle ; pour l'étude de la variabilité spatiale au sein de parcelles culturales de grandes dimensions, plutôt que de multiplier les fosses d'observation, on pourrait utiliser des cartes pénétrométriques (Billot, 1982), d'obtention plus rapide ; ceci impliquerait d'avoir calé les mesures sur les données morphologiques, en tenant compte de la texture et de l'état hydrique.

D'un autre côté, les conséquences des dégradations physiques mises en évidence doivent être précisées. Qu'il s'agisse des effets sur les paramètres de transfert, sur la pullulation de certains pathogènes ou prédateurs telluriques ou sur l'enracinement, les études à plusieurs échelles, associant différentes approches qualitatives et quantitatives, semblent prometteuses.

REFERENCES

Billot, J.F. 1982. Les applications agronomiques de la pénétrométrie à l'étude de la structure des sols travaillés. Sc. Sol, 3, 187-202.

Boiffin, J. 1984. La dégradation structurale des couches superficielles du sol sous l'action des pluies. Thèse DDI, INA-PG Paris.

Fies, J.C. 1971. Recherche d'une interprétation texturale de la porosité des sols. Ann. Agron., 22(6), 655-686.

Fies, J.C. & Stengel, P. 1981. Densité texturale des sols naturels : I - Méthodes de mesure. II - Eléments d'interprétation. Agronomie, 1(8), 651-666.

Guérif, J. 1982. Compactage d'un massif d'agrégats : effet de la teneur en eau et de pression appliquée. Agronomie, 2(3), 287-294.

Guérif, J. 1984. The influence of Water-content Gradient and Structure Anisotropy on Soil Compressibility. J. Agric. Engng. Res., 29, 367-374.

Hénin, S., Gras, R., Monnier, G. 1969. Le profil cultural. 2ème Edit. (Masson, Paris).

Hodgson, J.M. 1974. Soil Survey Field Handbook. Tech. Monog. N°5 (Soil Survey, London).

Manichon, H. 1982a. Influence des systèmes de culture sur le profil cultural : élaboration d'une méthode de diagnostic basée sur l'observation morphologique. Thèse DDI, INA-PG Paris.

Manichon, H. 1982b. L'action des outils sur le sol : appréciation de leurs effets par la méthode du profil cultural. Sc. Sol, 3, 203-219.

Manichon, H., & Sebillotte, M. 1973. Etude de la monoculture du Maïs ; résultats d'enquêtes agronomiques en Béarn. Doc. mult. Chaire d'Agronomie INA-PG (Paris).

Manichon, H., & Sebillotte, M. 1975. Analyse et prévision des conséquences des passages successifs d'outils sur le profil cultural. Bull. Tech. Inf. (Ministère de l'Agriculture), N°302-303, 569-577.

Manichon, H., & Bodet, J.M. 1976. Incidence de la simplification du travail du sol sur l'évolution des profils culturaux. In "Simplification du travail du sol en culture céréalière" (ITCF, Paris).

Manichon, H., Leterme, Ph., Buisson, O. 1985. Profils culturaux et enracinements. Doc. mult. INRA-ONIC, Paris.

Monnier, G., & Stengel, P. 1982. Structure et état physique du sol. Encyclopédie des Techniques Agricoles (Paris).

Nikiforoff, A. 1941. Morphological classification of soil structure. Soil Sc., 52, 193-211.

Papy, F. 1984. Comportement du sol sous l'action des façons de reprise d'un labour de printemps : effets des conditions climatiques et de l'état structural. Thèse DDI, INA-PG Paris.

Sebillotte, M. 1978. Itinéraires techniques et évolution de la pensée agronomique. C.R. Acad. Agric. Fr., 906-913.

Soane, B.D., Campbell, D.J., Herkes, S.M. 1971. Hand held gamma ray transmission equipment for the measurement of bulk density of field soils. J. Agric. Engng. Res., 16, 145-156.

Stengel, P. 1979. Utilisation de l'analyse des systèmes de porosité pour la caractérisation de l'état physique du sol in situ. Ann. Agron., 30(1), 27-51.

Tardieu, F. 1984. Influence du profil cultural sur l'enracinement du Maïs. Thèse DDI, INA-PG Paris.

Tardieu, F., & Manichon, H. 1986a. Caractérisation en tant que capteur d'eau de l'enracinement du Maïs en parcelle cultivée. I - Discussion des critères d'étude. Agronomie, 6(4).

Tardieu, F., & Manichon, H. 1986b. II - Une méthode d'étude de la répartition verticale et horizontale des racines. Agronomie, 6(5).

Comportement physique intrinsèque de mottes à macroporosité différente

P.Curmi

Institut National de la Recherche Agronomique, Laboratoire de Recherche associé à la Chaire de Science du Sol de l'Ecole Nationale Supérieure Agronomique de Rennes, France

RESUME

Dans le cadre de l'étude des rotations céréalières intensives, le comportement physique de mottes à forte macroporosité (non tassées) et de mottes à macroporosité réduite (tassées) a été établi sur deux matériaux différents : un limon moyen sableux et un limon argileux.

La teneur en eau pondérale et le volume apparent des mottes ont été mesurés aux faibles valeurs du potentiel hydrique (pF≤3)

A pF 1 ,la teneur en eau des mottes non tassées est significativement plus importante que celle des mottes tassées. Les mottes tassées sont, à ce pF, saturées en eau tandis que les mottes non tassées présentent encore une porosité libre à l'air.

INTRODUCTION

Les diverses opérations de travail du sol créent au sein de la couche labourée des états structuraux différents. Manichon (1982) a montré que ces états pouvaient être décrits à partir de l'abondance relative et de la distribution spatiale de deux types principaux de mottes (photo 1): des mottes non tassées présentant une forte macroporosité,qu'il appelle Γ, et des mottes tassées présentant une macroporosité réduite, qu'il dénomme Δ.Le comportement physique intrinsèque aux faibles succions de ces deux types de mottes, courbes de retrait et de rétention en eau, a été suivi au laboratoire au cours de la dessication sur deux matériaux de textures différentes.

MATERIELS ET METHODES

Des mottes à structure conservée, de 5 à 10 cm^3 de volume, sont extraites de l'horizon labouré non repris par les façons superficielles de deux parcelles expérimentales conduites en monoculture de blé.Sur le site du Rheu, cet horizon présente une texture limoneuse, sa fraction argileuse est constituée d'un mélange de quartz, kaolinite, illite, interstratifiés 1,0-1,4 nm et vermiculite hydroxyalumineuse. Sur le site de Grignon, sa texture est limono-argileuse et l'argile est un mélange de quartz,kaolinite, illite,interstratifiés 1,0-1,4 nm, vermiculite s.s. et smectites. Les compositions granulométriques et les teneurs en matières organiques sont présen-

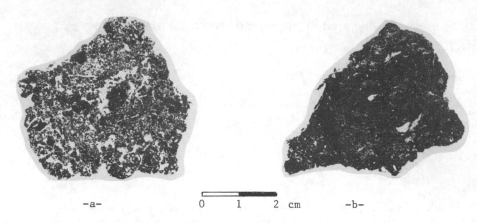

-a-   0   1   2 cm   -b-

Photo. 1   Macroporosité des deux types de mottes, (vides en blanc)
Coupes de mottes imprégnées par une résine fluorescente aux U.V.
-a- Motte non tassée, à forte macroporosité (Γ)
-b- Motte tassée, à macroporosité réduite (Δ)

TABLEAU 1   Caractéristiques physiques des deux matériaux

MATERIAU	Granulométrie en % de la terre fine					Matières organiques en %
	A	LF	LG	SF	SG	
LE RHEU	14.1	25.3	47.3	11.2	2.1	2.0
GRIGNON	22.1	22.2	43.0	10.0	1.9	2.5

tées dans le tableau 1.

Les mottes sont soumises à l'histoire hydrique suivante: à partir de leur humidité au moment du prélèvement (teneur en eau pondérale W de 18 % pour Le Rheu et de 13 % pour Grignon), elles sont réhumectées à pF 1 (P=0.01 bar), puis subissent une dessication par étapes jusqu'à pF 3 (P=1 bar) en utilisant le dispositif d'ultrafiltration de Tessier & Berrier (1979) pour contrôler le potentiel hydrique. Une dizaine de répétitions sont réalisées par point de mesure.

Pour chaque valeur de potentiel, la densité apparente des mottes est déterminée par poussée d'Archimède dans le pétrole (Monnier & al.,1973) et la teneur en eau par gravimétrie.

RESULTATS

Volume apparent

Les courbes de retrait (fig.1) montrent que les mottes limono-argileu-

54

se de Grignon subissent un retrait au cours de la dessication entre pF1 et
pF3, tandis que les mottes limoneuses du Rheu ne présentent pas d'évolution
significative.

Comme Manichon (1982) l'a montré, les valeurs de l'indice des vides e
sont plus élevées et présentent une plus forte dispersion pour les mottes à
forte macroporosité par rapport aux mottes à macroporosité réduite,confir-
mant le bien-fondé de la distinction morphologique des mottes en deux types
Γ et Δ .

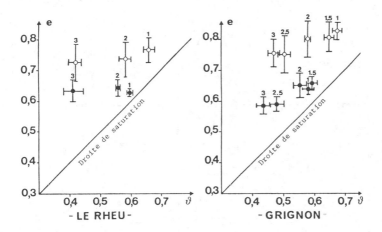

Fig. 1   Courbes de retrait des deux types de mottes.
O mottes Γ à forte macroporosité; ● mottes Δ à macroporosité réduite.
1,2,3... pF appliqué; ⊞ intervalle de confiance pour un risque α =0.05

Taux de saturation de la porosité et courbes de rétention en eau

En comparant, en fonction du potentiel hydrique, l'évolution du volume
apparent des mottes représenté par l'indice des vides e et la courbe de rétention en eau représentée par l'indice d'eau ϑ , on constate que les mottes à macroporosité réduite Δ présentent un taux de saturation de la porosité θ_S plus élevé et sur une gamme de pF plus étendue que les mottes à
forte macroporosité Γ (fig.2 et tab.2). Ainsi à pF 1, θ_S est supérieur à
90 % et la porosité reste saturée à plus de 80% jusqu'à pF 2.5 pour les
mottes Δ, tandis que pour les mottes Γ, θ_S est compris entre 82 et 86% à
pF 1 et la porosité reste saturée à plus de 80% jusqu'à pF 2 pour le site
du Rheu et pF 1.5 pour le site de Grignon.

En fonction de la texture, le taux de saturation à pF 1 est plus élevé de 4% sur le site limoneux du Rheu par rapport au site limono-argileux
de Grignon pour les deux types de mottes.

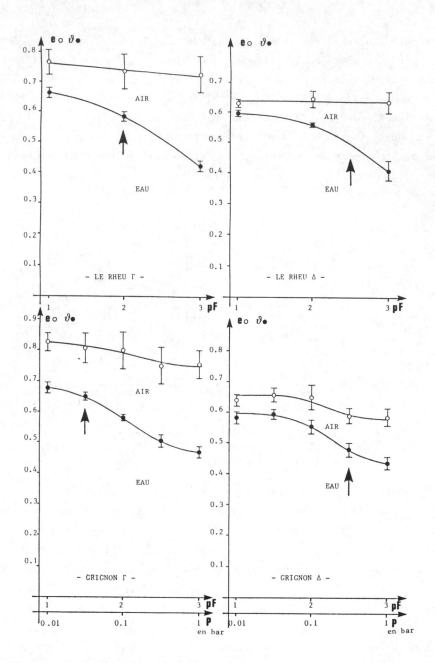

Fig. 2   Evolution du volume apparent et courbes de rétention
en eau des mottes en fonction du potentiel hydrique.

↑ pF correspondant à un taux de saturation en eau de la porosité de 80 % .

TABLEAU 2   Taux de saturation en eau de la porosité des mottes

$$\theta_S = (\vartheta/e) \times 100$$

pF	LE RHEU		GRIGNON	
	Γ	Δ	Γ	Δ
1.0	86.6 ±4.5	94.7 ±2.0	82.1 ±3.3	91.0 ±3.1
1.5	–	–	80.6 ±4.6	90.3 ±3.0
2.0	79.5 ±5.8	86.6 ±3.1	72.7 ±5.3	85.2 ±4.9
2.5	–	–	67.2 ±5.5	81.0 ±3.8
3.0	58.2 ±5.3	65.1 ±4.9	62.2 ±4.3	74.5 ±3.9

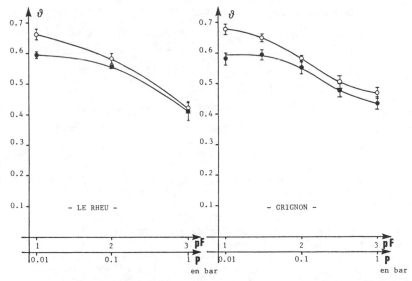

Fig. 3   Courbes de rétention en eau des deux types de mottes.
○ Mottes à forte macroporosité Γ
● Mottes à macroporosité réduite Δ

Si l'on compare enfin les courbes de rétention en eau pour les deux types de mottes (fig. 3 et tab. 3), on constate que l'indice d'eau à pF 1 est plus élevé pour les mottes Γ que pour les mottes Δ, alors que ces mottes Γ ont un taux de saturation de la porosité plus faible. Cette différence se réduit au cours de la dessication. Ces différents résultats confirment les travaux de Papy (1984).

TABLEAU 3 Accroissement de la teneur en eau pondérale lorsque la macroporosité des mottes croît: $\dfrac{(\vartheta_\Gamma - \vartheta_\Delta)}{\vartheta_\Delta} \times 100$

pF	LE RHEU	GRIGNON
1.0	11.4 ±2.6	16.5 ±3.4
1.5	-	9.5 ±2.6
2.0	4.5 ±2.1	5.0 ±3.0
2.5	-	n.s
3.0	n.s	7.6 ±3.9

CONCLUSION

Le suivi simultané du volume apparent et de la teneur en eau, en contrôlant le potentiel hydrique, a permis de mettre en évidence le rôle de l'organisation du sol et plus particulièrement de sa macroporosité sur la rétention de l'eau au faibles potentiels. La diminution de la macroporosité sous l'effet des façons culturales conduit, aux faibles potentiels de l'eau (pF<3), à des risques d'anoxie (θ_s>80%). Ces risques sont plus élevés pour les matériaux limoneux que pour les matériaux limono-argileux.

REFERENCES

Manichon,H. 1982. Influence des systèmes de culture sur le profil cultural: élaboration d'une méthode de diagnostic basée sur l'observation morphologique. Thèse Doct. Ing. INAPG, Paris, 214p.

Monnier, G. Stengel, P. & Fies, J.C. 1973. Une méthode de mesure de la densité apparente de petits agglomérats terreux. Application à l'analyse des systèmes de porosité du sol. Ann. Agron. 24 (5), 533-545.

Papy, F. 1984. Comportement du sol sous l'action des façons de reprise d'un labour au printemps. Effets des conditions climatiques et de l'état structural. Thèse Doct. Ing. INAPG, Paris, 232p.

Tessier, D. & Berrier, J. 1979. Utilisation de la microscopie électronique à balayage dans l'étude des sols. Observations de sols humides soumis à différents pF. Bull. AFES, Science du Sol, 1 : 67-82.

Pore size distribution of soils as determined from soil characteristic curves using non-polar liquid

S.Aggelides

Soil Science Institute of Athens, Lycourissi, Attiki, Greece

ABSTRACT

Moisture characteristic curves for inert porous materials have been used for assessing their pore size distribution while characteristics for shrinking and swelling materials with non-polar liquid have been obtained for the same purpose. Changes of the pore size distribution in subsequent cycles of wetting and drying with water have also been used for assessing structural stability in soils. Here we describe a procedure for studying the total effect of absorption and desorption of water on pore size distribution of a soil and subsequently the stability of soil structure. For this, characteristic curves with a non-polar liquid (alcohol) have been obtained for assessing the initial soil structure and followed by moisture characteristic curves for cycles of drying and wetting.

INTRODUCTION

Pore size distribution of inert porous materials can be assessed from soil moisture characteristic curves and it can be used as a measure to evaluate their structure (Childs 1940, Leamer and Lutz, 1940, Vomocil, 1965). Childs (1940) used differential curves of moisture characteristics obtained during subsequent cycles of drying and wetting to assess changes of the pore space of soil consisted of clay aggregates and used them as a measure of stability of the aggregates. He proposed that the ratio of the height of the peaks of successive differential curves can be used as a measure for soil stability. Childs and Youngs (1959) used moisture characteristic curves for assessing the soil stability of opencast coal sites. Moisture characteristics however do not allow the assessing of the total changes of pore space due to wetting, since, during the first wetting serious alterations ofthe pore space take place in non inert porous materials. These alterations are due to swelling of aggregates as well as to their disruptions.

The technique based on forcing mercury into the pores is not very accurate since it is applied to very small samples (Klock et al. 1969).

In the present work a procedure is used for assessing total changes of the porous space due to wetting with water and subsequently for estimating the stability of structure. For this a drying characteristic curve with a

59

non-polar liquid is obtained initially and followed by moisture characteristics obtained during subsequent cycles of wetting and drying.

MATERIALS AND METHODS

The soil characteristic curves were obtained by the Haines' Buchner funnel method (Haines, 1930). The sintered funnel was connected to a burette through a flexible tube which could be lowered or ascented. The negative pressure head imposed to the sample was measured by the distance between the free liquid level in the burette and the middle height of the sample. The sintered funnel and burette were covered for minimising evaporation losses. Soil samples of a volume of 40 cm^3 were embeded in a cylindrical ring of a height of 3 cm and an internal diameter 5,3 cm and put in contact with the sinter of the funnel. The bottom of the ring consisted of cotton cloth. The gap left between funnel and ring was filled with paraffin wax to avoid storage of water or ethyl alcohol during saturation of the same. Some properties of the soils used are shown in table 1.

TABLE 1   Soils used and some of their properties

Material	Clay	Silt	Sand	Organic matter	$CaCO_3$
	%	%	%	%	%
Sand					
Xyniada soil	24	30	46	10.2	0
Chalandri soil	21	25	54	2.1	43
Kopaida soil	51	27	22	5.1	41

Soil samples thus prepared were first wetted with ethyl alcohol and their drying characteristic curves were obtained by increasing step by step the pressure head applied. Subsequently the soil samples were removed from the funnels and left to air-dry; they were placed again in the funnels, wetted with water and drained by increasing the pressure head applied. Both alcohol and water used were boiled to remove dissolved air. The pore size distribution was obtained from the soil characteristic curves by graphical estimation of their slopes as a function of the negative pressure. In order to compare the pore size distribution as determined from the characteristic curves obtained with water and alcohol, the latter curves were transformed to the equivalent moisture characteristics. This was done

60

by expressing pressures in alcohol in pressures in water by using equation (1).

$$P_w = P_\alpha \left[\frac{\gamma_w \cos\theta_w}{\gamma_a \cos\theta_a} \right]$$ (1)

where P, γ and θ represent the pressure, the surface tension coefficient of the liquid and the contact angle respectively. The subscripts w and a refer to water and alcohol respectively. Equation (1) relates the pressures P_w and P_a which are necessary to drain a pore of diameter, say D. The quantities $\gamma_w \cos\theta_w$ and $\gamma_a \cos\theta_a$ were obtained independently after the experiment was completed by the capillary rise of the liquid drained from the soil samples in a capillary tube of known diameter, d, using the equation,

$$\gamma \cos\theta = \frac{d\rho g h}{4}$$ (2)

where h is the height to which the liquid was risen, ρ is the density of the liquid, which was also measured and g is the acceleration due to gravity. Table 2 shows the measured values of $\gamma_w \cos\theta$ and $\gamma_a \cos\theta_a$ for each experiment together with the temperature during measurements.

TABLE 2

Material	$\gamma_w \cos\theta_w$ dynes/cm	$\gamma_a \cos\theta_a$ dynes/cm	Temperature
Sand	61.36	23.60	24±2°C
Xyniada soil	63.11	22.61	19±2°C
Chalandri soil	55.43	22.53	19±2°C
Kopaida soil	53.60	24.31	24±2°C

In order to assess the workability of the procedure, characteristic curve with alcohol of a sand was transformed to moisture characteristic and compared to the moisture characteristic obtained experimentally (see fig. 1). It is seen from fig. 1, that the pore size distribution for the inert material, as revealed by both procedures lie very close. One may add here that successive draining characteristics with alcohol for soil aggregates did not differ substantially.

RESULTS AND DISCUSSION

In figures 2 and 3 characteristic curves of soil aggregates of Xyniada with their respective differential curves are shown. Two soil samples consisted of aggregates of a size between 1-2 mm were air-dried (volumetric water content 8%) and were uniformly packed into two rings. The characteri-

61

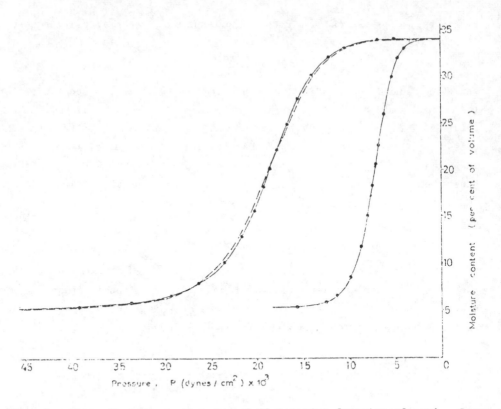

Fig. 1 - Characteristic curves in a 0.25-1mm size fraction of sand. Curve
(1) obtained by using ethyl alcohol. Curve (2) obtained by using water.
Curve (3) equivalent moisture characteristic of curve (1).

stic curves were obtained by using alcohol in one sample and water in the
other. The water equivalent characteristic curve of the characteristic
curve with alcohol was determined, by using the horizontal axis
transformation (eq. (1)) and the resultant curve was plotted as curve 1
(fig. 2). Curve 2 represents the characteristic curve obtained by using
water. The samples were then air dried and saturated again with alcohol
and water respectively as it was done in the first cycle of the experiment.
The characteristic curve obtained by using alcohol at the second cycle was
identical to that of the first. Curve 3 represents the moisture
characteristic curve referring to the second cycle of wetting. Curves 1´,
2´ and 3´ in fig. 3 represent the differential curves of 1, 2 and 3 of fig.
2. As one can see the initial pore size distribution, as this is revealed
by curve 1´, is moved to the left. Some of the large pores emptying at a
pressure $P_1 = -7.2 \times 10$ dynes/cm² have been destroyed after the first cycle of

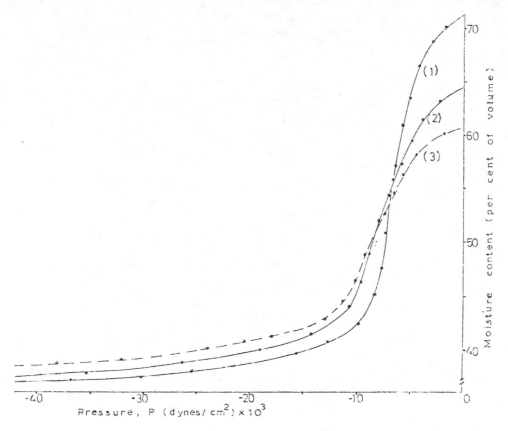

Fig. 2 – Characteristic curves for 1-2 mm aggregate fraction from 0-30 cm layer, Xyniada soil. Curve (1) moisture equivalent characteristic curve obtained by using alcohol. Curve (2) moisture characteristic after first wetting of the aggregates. Curve (3) moisture characteristic after second wetting of the aggregate

wetting; giving way to some smaller pores, gathered around a pressure $P_2 = -8.6 \times 10^3$ dynes/cm². Some of the latter pores have been also destroyed during the second cycle of wetting to giving way to another group of even smaller pores, gathered at a pressure $P_3 = -10.0 \times 10^3$ dynes.cm⁻². The effect of successive cycles of wetting and drying to the pore size distribution of soil aggregates was investigated first by Childs (1940) and a similar behaviour has been reported. In the present case the stability factor according to Childs would have been equal to the ratio of the peaks $h_3/h_2 = 2.50/3.25$, where h_3 and h_2 represent the peak values of the curves 3′ and 2′ respectively. In a similar way using the peak of curve 1′ the

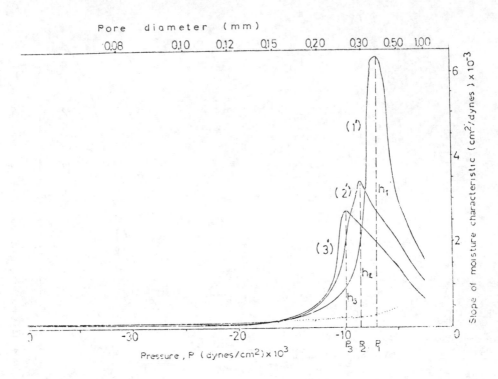

Fig. 3 – Differential curves of the characteristic curves in fig.2. Curves (1′), (2′) and (3′) derived from curves (1), (2) and (3) respectively.

ratio h_2/h_1 = 3.25/6.00 would indicate the effect of water after the first wetting of the aggregates and the ratio h_3/h_1 = 2.5/6.00 would indicate the effect of water after the second wetting. In figures 4 and 5 characteristic curves of the soil from Chalandri with their differential curves are shown. In this case the sample was undisturbed and it was taken from the surface layer some days after tillage. The sample was saturated with alcohol and the characteristic curve was obtained. Its water equivalent characteristic curve was plotted as curve 1. Then it was air dried and saturated again, this time with water and the moisture characteristic curve was obtained as curve 2 (fig. 4). Curves 1′ and 2′ in fig. 5 represent the differential curves of 1 and 2 of fig. 4. Comparison between the pore size distribution before and after the action of water shows that the big pores are reduced giving way to smaller pores.

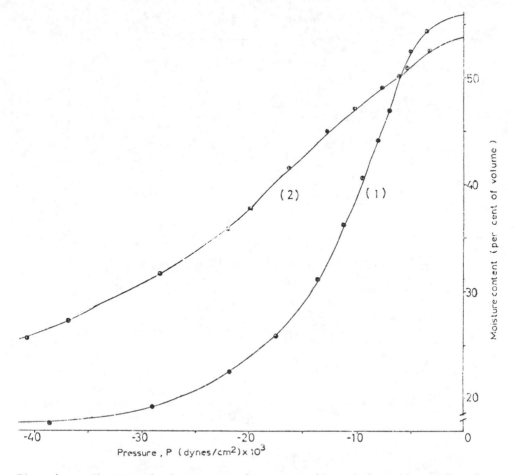

Fig. 4 – Characteristic curves for an undisturbed soil sample from Chalandri. Curve (1) moisture equivalent characteristic curve obtained by using ethyl alcohol. Curve (2) moisture characteristic curve obtained experimentally.

In figures 6 and 7 are shown characteristics and their differential curves for a sample from a clay soil. The difference in liquid content at saturation can be attributed to the swelling of the sample causing an increase of the total porous space.

From the results shown one may conclude that the procedure applied here can given information about the total effect of wetting with water on the pore space of soils.

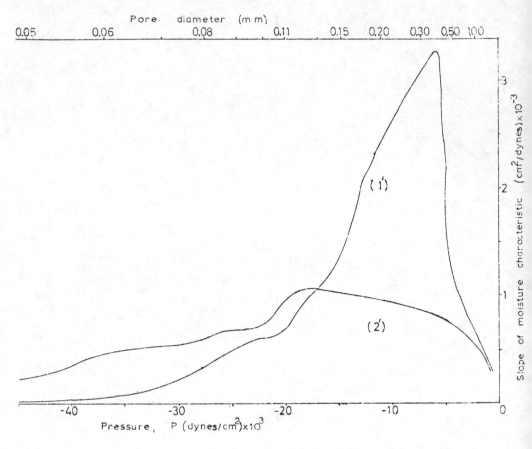

Fig. 5. Differential curves of the characteristic curves in fig. 4. Curves (1´) and (2´) derived from curves (1) and (2) respectively.

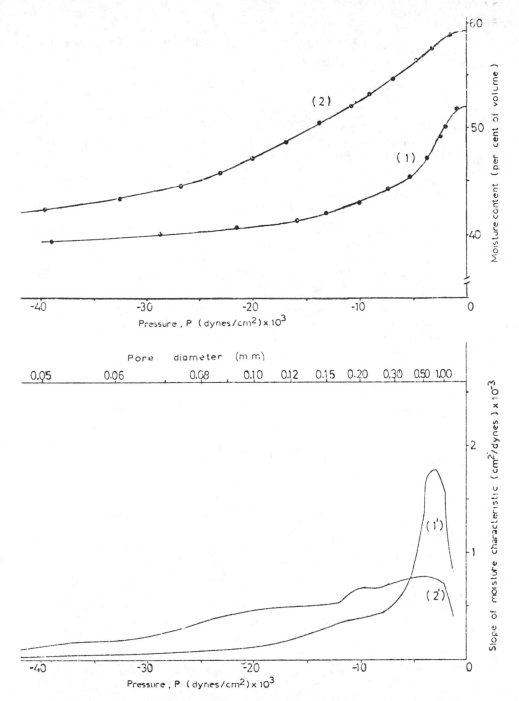

Figs. 6, 7. Characteristic curves of a clay soil and their differentials. Curve (1) moisture equivalent characteristic curve with alcohol. Curve (2) moisture characteristic curve. Curves (1′) and (2′) derived from curves (1) and (2) respectively.

ACKNOWLEDGEMENTS

Thanks are extended to Mrs. Chr. Kotsia for valuable assistance during the execution of the experiments.

REFERENCES

Childs, E.C. 1940. The use of the soil moisture characteristic in soil studies. Soil Sci. 50, 239-52.

Childs, E.C. and Youngs, E.G. 1959. The use of moisture characteristic curves in assessing the soil stability of opencase coal sites. International symposium on soil structure, Ghent 1958. Overdruk nit mededelingen van de landbouwhogeschool en de opzoekingsstation van de staat de Gent. Vol. 24, 415-21.

Haines, W.B. 1930. Studies in the physical properties of soil. Jour. Agr. Sci. 20, 97-116.

Klock, G.O., Boersma, L. and De Backer, L.W. 1969. Pore size distribution as measured by the mercury intrusion method and their use in predicting permeability. Soil Sci. Soc. Amer. Proc. 33, 12-15.

Leamer, R.W. and Lutz, J.F. 1940. Determination of pore size distribution in soils. Soil Sci. 49, 347-361.

Vomocil, J.A. 1965. Porosity. In C.A. Black, ed. Methods of soil analysis, Agronomy 9. Part 1. Physical and mineralogical properties, including statistics of measurement and sampling. Madison, Wisconsin, American Society of Agronomy. pp. 299-314.

A model to predict subsoil compaction due to field traffic

J.J.H.van den Akker & A.L.M.van Wijk

Institute for Land and Water Management Research (ICW), Wageningen, Netherlands

ABSTRACT

Intensification of crop rotation and advancing mechanization are ever more coupled with compaction of the subsoil. In addition to field studies there is an increasing need for tools to evaluate effects of field traffic on soils at variable conditions of texture, density and moistening.

This paper describes the principle and application of a first version of a model that computes compaction of the subsoil due to field traffic. The model is based on the work of FRÖHLICH (1934) who started from the theory of Boussinesq for a point load on a isotropic semi-infinite solid mass. Starting from the horizontal and vertical stresses under wheels acting on the subsoil the model calculates the horizontal, vertical and shear stresses in a 3-dimensional space in the subsoil. From these results the principal and the mean stresses are computed. A concentration factor is incorporated in the model to account for the non-pure elastic behaviour of soils. On the basis of relationships between stress and bulk density for a particular soil, obtained from triaxial tests, the calculated stress distribution can be converted into a density distribution.

The model considers subsoil compaction only, because the model theory permits merely small deformations.

1. INTRODUCTION

During the last 10-15 years general surveys in the field as well as complaints from practice reveal considerable soil degradation occurring as deterioration of soil structure and an increase of compaction of the subsoil. In the same time two developments could be observed in agriculture in the Netherlands: i) an advancing mechanization with increasing wheel loads and ii) an intensification of crop rotation. To a great extent cereals have been replaced by root crops and forage maize, producing a considerably greater mass to harvest at wetter soil conditions. At this moment more than 60% of arable land in the Netherlands is used for growing potatoes (23%), sugar beets (18%) and forage maize (22%). To reduce the negative consequences of mechanization wheel equipments are adapted mainly by applying wider tires. In spite of that subsoil compaction becomes ever more serious and extends to a depth beyond that of the annual main soil tillage, just on account of the wider tire sizes.

The extent to which a soil will be compacted during field operations depends on the type of field traffic as well as on the prevailing soil conditions. The pressures exerted to the soil by wheel traffic may very widely depending on wheel properties (dimension, ply rating and inflation pressure of tires) and wheel load. Moreover, when loaded, soils behave differently depending on the load, soil texture, bulk density and moisture conditions. However, it is not attainable to study the effects of all possible combinations of these factors in the field. Also the transferability of data obtained from experimental fields to other combinations of soil and traffic is often rather difficult. Therefore in addition to field studies there is a great need for models, using both traffic and soil characteristics as input data, that enable to study soil compaction at a wide variety of traffic and soil conditions.

This paper presents a first version of a model that predicts compaction of the subsoil as result of horizontal and vertical forces originating from wheel traffic. The first results of the model calculation are principal and mean stress distributions in the subsoil. With the aid of relationships between these stresses and bulk density obtained from tri-axial tests the stress distribution can be converted into a density distribution. To verify the model results and to derive the stresses in the tire-soil contact surface controlled traffic experiments have been executed in the field. With the aid of markers placed previously in a grid perpendicular to the direction of moving of the wheel the horizontal and vertical displacement of the

70

soil under the wheel have been established. From this displacement the change in bulk density can be calculated, which is also measured by core sampling.

2. PRINCIPLE OF THE MODEL

The model is based on the theory of Boussinesq (1885) describing the distribution of stresses in a homogeneous isotropic semi-infinite solid mass due to a force applied on a point on the surface of that mass. On any volume element in the semi-infinite solid vertical, horizontal and tangential normal stresses and vertical and horizontal shear stresses are operative (Fig. 1). For the different stresses can be written:

$$\sigma_z = \frac{3P}{2\pi r^2} \cos^3 \theta$$

$$\sigma_h = \frac{P}{2\pi r^2} \left\{ 3 \cos \theta \sin^2 \theta - \frac{m-2}{m} \frac{1}{1 + \cos \theta} \right\}$$

$$\sigma_t = -\frac{m-2}{m} \frac{P}{2\pi r^2} \left\{ \cos \theta - \frac{1}{1 + \cos \theta} \right\}$$

$$\tau_z = \frac{3P}{2\pi r^2} \cos^2 \theta \sin \theta = \tau_h$$

(1)

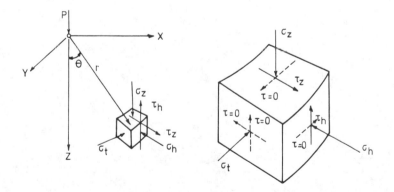

Fig. 1. Composition of stressec acting in a volume element due to a point load on a semi-infinite solid

where:

σ_z, σ_h, σ_t = vertical, horizontal and tangential stress respectively

τ_z, τ_h = vertical and horizontal shear stress

P = vertical point load

r and θ = polar coordinates

m = inverse of the Poissons' ratio.

To account for symmetry τ_z and $\tau_h \neq 0$. The factor m gives the ratio between the deformation parallel (ε_1) and perpendicular (ε_2) to the direction of the uni-axial loading of an element. So:

$$m = \frac{\varepsilon_1}{\varepsilon_2} \qquad (2)$$

m may have values between 2 and 3. No change of volume of the soil occurs when m = 2.

Since the inaccuracy of results is small when m $\neq$ 2 FRÖHLICH (1934) proposed to set m = 2 in the eqs. (1). Moreover, he introduced a concentration factor (ν) in the Boussinesq's formulas to account for the non-purely elastic behaviour of soils. The compressive stress in the soil tends to concentrate around the load axis. This behaviour becomes more pronounced when the soil is more plastic due to greater wetness. The concentration factor expresses the rate of distribution of loading around the load axis. Inserting m = 2 and ν in the eqs. (1) gives:

$$\sigma_z = \frac{\nu P}{2 \pi r^2} \cos^\nu \theta$$

$$\sigma_h = \frac{\nu P}{2 \pi r^2} \cos^{\nu-2} \theta \sin^2 \theta$$

$$\sigma_t = 0 \qquad (3)$$

$$\tau = \frac{\nu P}{2 \pi r^2} \cos^{\nu-1} \theta \sin \theta$$

When ν = 3 the formulas of Boussinesq are obtained with m = 2. The value of ν is greater when the soil is more soft. KOOLEN and KUIPERS

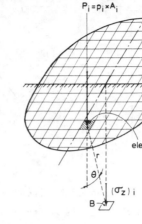

$P_i = p_i \times A_i$

element i

r

θ

$(\sigma_z)_i$

B

Fig. 2. Procedure of summation of
partial stresses σ_{z_i} originating from point loads P_i
distributed over the
wheel-soil contact surface
(after SÖHNE, 1953)

(1983) give values of ν of 3, 4 and 5 for a hard, normal and soft soil
respectively.

SÖHNE (1953) developed a numerical procedure for calculating the
vertical stresses σ_z in the soil due to loading by a tire. A tire does
not transfer its load to a singel point but to the entire soil-tire
contact surface. To account for this Söhne divided the contact area in
a number of elements in whose centres a given point load acts (Fig. 2).
The vertical stress σ_z in a particular point B in the soil is obtained
by adding all the vertical stresses exerted by all these point loads.
However, this procedure cannot be applied to calculate the horizontal
stresses σ_h and shear stresses τ_z. The directions of these stresses
acting in a given point in the soil and exerted by different point
loads are different since the different orientation of the point loads
to that point in the soil. Therefore these stresses may not be simply
added to obtain the total σ_h and τ_z acting in a given point in the
soil. Addition of these stresses becomes possible when they are decomposed in the x- and y-components. This is illustrated in Fig. 3.
TIMOSHENKO and GOODIER (1980) give the following solution for this
decomposition:

$$\sigma_x = \sigma_h \cos^2 \phi + \sigma_t \sin^2 \phi - 2\tau_{ht} \sin \phi \cos \phi$$

$$\sigma_y = \sigma_h \sin^2 \phi + \sigma_t \cos^2 \phi + 2\tau_{ht} \sin \phi \cos \phi \qquad (4)$$

$$\tau_{xy} = (\sigma_h - \sigma_t) \sin \phi \cos \phi + \tau_{ht} (\cos^2 \phi - \sin^2 \phi)$$

73

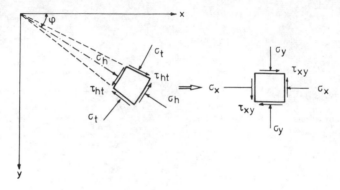

Fig. 3. Decomposition of horizontal and shear stresses in x- and y-
components in aid of the addition of these stresses

Since $\tau_{ht} = \sigma_t = 0$ eqs. (3) can be rewritten as:

$$\sigma_x = \sigma_h \cos^2 \phi$$

$$\sigma_y = \sigma_h \sin^2 \phi \qquad\qquad (5)$$

$$\tau_{xy} = \sigma_h \sin \phi \cos \phi$$

The shear stress τ_z in the z-plane can be decomposed in:

$$\tau_{zx} = \tau_{xz} = \tau_z \cos \phi$$

$$\tau_{zy} = \tau_{yz} = \tau_z \sin \phi \qquad\qquad (6)$$

The symbols used in eqs. (4, 5 and 6) are explained in Fig. 3.

FRÖHLICH (1934) indicates how stresses in the soil exerted by a
single horizontal point load can be calculated. This approach is
applied to calculate stresses originating from the horizontal point
loads acting in the centres of the same elements in the tire-soil contact surface where also the vertical point loads are acting (see
Fig. 2). Thus the stresses σ_x, σ_y, σ_z, τ_{xy}, τ_{xz} and τ_{yz} can be computed in any point in the soil for each given vertical and horizontal
load. From these stresses the principal stresses S can be derived.
After TIMOSHENKO and GOODIER (1980) these principal stresses are the
roots (S_1, S_2 and S_3) of the following equation:

$$S^3 - (\sigma_x + \sigma_y + \sigma_z)S^2 + (\sigma_x\sigma_y + \sigma_y\sigma_z + \sigma_x\sigma_z - \tau_{xy}^2 - \tau_{xz}^2 -$$

$$- \tau_{yz}^2)S - (\sigma_x\sigma_y\sigma_z + 2\tau_{xy}\tau_{xz}\tau_{yz} - \sigma_x\tau_{yz}^2 - \sigma_y\tau_{xz}^2 - \sigma_z\tau_{xy}^2) = 0$$

$$(7)$$

It will be clear that these very detailed calculations can only
be done by computer. The present computer model is capable to calculate all stresses in intersections under the wheel both parallel and
perpendicular to the direction of moving due to vertical as well as
horizontal pressures.

The theory used in the model is properly speaking not applicable
to describe plastic deformation and does not permit great volume
changes. In praxis the deformation and compaction of the generally
loose topsoil due to wheel loads will be often so much that it is not
permitted to use the model under those conditions. However, the problem of soil compaction concerns mainly the subsoil, since a too high
subsoil density hampers root growth. A shallow rooting depth restricts
the water supply from the subsoil and groundwater table. Moreover, as
opposed to the topsoil, compaction of the subsoil can only be remedied
with great effort. The deformation and volume change occurring in the
subsoil will not be very great generally. To predict subsoil compaction the model calculation starts from the vertical and horizontal
pressures occurring in a plane under the wheel located at the base of
the topsoil.

3. CONTROLLING FIELD EXPERIMENTS

In cooperation with the Research Station for Arable Farming and
Field Production of Vegetables, Institute of Agricultural Engineering
and the Soil Tillage Laboratory of the Agricultural University controlled traffic experiments have been carried out in threefold with
a single wheel tester on a sandy soil to verify model results. The
wheel load applied amounted to 32 kN. The tire was a Vredestein
Special Ribbed 16.0/70-20 with an inflation pressure of 2.4 bar. The
following measurements have been made: displacement of soil elements
and bulk density distribution with depth, pulling power to move on the
single wheel tester and the vertical stresses exerted by the loaded

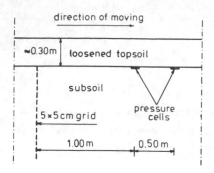

Fig. 4. Field experimental set up
to measure vertical
stresses and displacement
of soil elements under
wheels

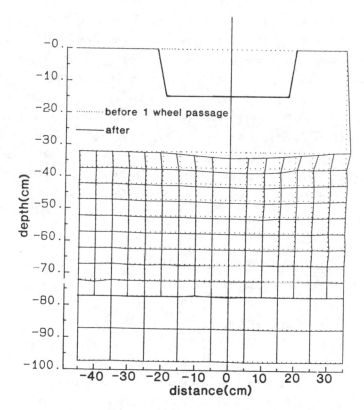

Fig. 5. Measured deformation of a grid in the subsoil under the rut,
perpendicular to the direction of moving

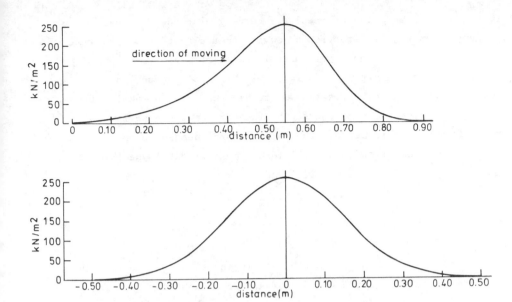

Fig. 6. Vertical stresses, parallel (upper part) and perpendicular
(lower part) to the axis of the rut, acting on a depth of 30
cm during passing of a 16.0/70-20 ribbed tire, loaded to 32 kN
and inflated to 2.4 bar

topsoil upon the subsoil. The experimental set up is given in Fig. 4.

To measure the vertical stresses two pressure cells were placed
on a mutual distance of .50 cm in the centre-line of the rut. To regis-
ter the horizontal and vertical displacement of the soil under the
wheel, markers were placed previously in a 100 cm wide grid of 5 x 5 cm
between 30 and 105 cm depth, perpendicular to the direction of moving
and at a distance of 1 m before the first pressure cell. To realize
homogeneous conditions in the topsoil, the topsoil was removed over a
width of 1.5 m starting 1 m before the grid to 1 m beyond the second
pressure cell. After installation of the pressure cells the topsoil
was replaced in two layers which were lightly compacted by rolling.

In Fig. 5 the deformation of the subsoil is presented in a grid
perpendicular to the direction of moving. As an average of two pressure
cells Fig. 6 shows the vertical stresses both parallel and perpendicular to the axis of the rut, acting on a depth of 30 cm during passing
of the wheel. The upper curve is entirely based on pressure cell read-

ings. The maximum value of the lower curve giving the vertical stresses perpendicular to the direction of moving equals to that of the upper curve. The shape of the lower curve is based on measurements during controlled traffic experiments in a soil bin and the fact that the total sum of vertical stresses has to be equal to the wheel load.

The horizontal force exerted by the tire upon the soil was derived from the ratio between the pulling force needed to move on the single wheel tester and his wheel load. The horizontal wheel load amounted to 26.4% of the vertical load.

4. COMPUTATION OF STRESSES IN THE SUBSOIL

For each of the 5 x 5 cm elements in the tire-soil contact surface projected to a depth of 30 cm (see Fig. 2) the vertical point load can be derived from the 3-dimensional vertical stress distribution described by the two mutually perpendicular stress distributions shown in Fig. 6. To account for the horizontal force exerted by the

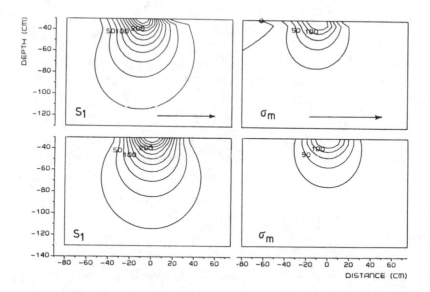

Fig. 7. Computed principle stress (S_1) and mean normal stress (σ_m) distributions in kN.m^{-2} parallel (upper part) and perpendicular (lower part) to the direction of moving

tire upon the soil a horizontal shear stress of 26.4% of the size of the vertical stress was applied. That means that the distribution of horizontal shear stresses is supposed to be proportional to the vertical stress distribution.

For the concentration factor ν in eqs. (3) a value of 4 was taken.

To compute the stresses a 5 x 5 cm grid with a width of 1.50 m is laid over the subsoil under the wheel to a depth of 1.0 m. In each nodal point of two intersections, the one parallel and the other through the axis of the wheel perpendicular to the direction of moving, the vertical (σ_z), horizontal shear (σ_z, σ_y), shear $(\tau_{xy}, \tau_{xz}, \tau_{yz})$, and principal stresses (S_1, S_2, S_3), yielding finally the mean normal stress (σ_m), are calculated in the way as indicated in Section 2. Fig. 7 shows some examples of these computations.

5. RELATIONSHIP BETWEEN STRESSES AND COMPACTION

Subsoil compaction is primarily observed in the Netherlands on light sandy loam and sandy soils. In literature stress-compaction relationships are given but mainly for other soil types. In civil engineering much attention is paid to the compaction behaviour of sands. In case of a pure-elastic isotropic material the volume change ε_v depends on the mean normal stress $\sigma_m = \frac{1}{3}(S_1 + S_2 + S_3)$ only. The distortion is determined by the deviator stress $(\sigma_1 - \sigma_3)$.

From tri-axial tests with sand it appeared that the volume change ε_v is composed of a part determined by the mean normal stress σ_m and a part that depends on the deviator stress:

$$\varepsilon_r = \varepsilon_{vc} + \varepsilon_{vd}$$

with: $\varepsilon_{vc} = f(\sigma_m)$ \hfill (8)

$$\varepsilon_{vd} = f(S_1 - S_3)$$

If the distortion in the subsoil is neglectable, compaction of the subsoil is mainly caused by the mean normal stress σ_m.

The volume change due to the mean normal stress can be deter-

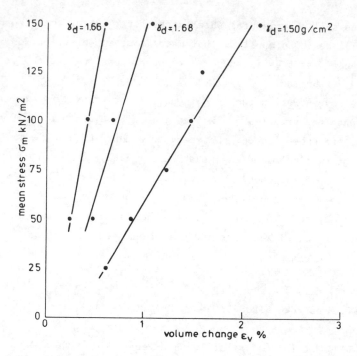

Fig. 8. Relationships between mean normal stress (σ_m) and volume
change (ε_v) for a medium fine sand at three bulk density
levels (γ_d)

mined rather simply by compacting soil samples under isotropic pressure.
In that case holds:

$$\sigma_m = S_1 = S_2 = S_3 \tag{9}$$

Fig. 8 shows relationships between σ_m and ε_v for the medium fine sandy
subsoil of the experimental field where the controlled traffic experiment described in Section 3 was carried out. The soil moisture content of the sand during the tri-axial test amounted to 11% (w/w),
equal to that during the traffic experiment. The sand contains 1.2%
organic matter and has the following particle size distribution:
3% < 2, 21% between 2-50, 40% between 50-150 and 36% > 150 μm.

When plastic deformation occurs the compaction will increase in
addition to the elastic part due to a plastic contribution. From the

80

computed principal stresses S_1 and S_3 can be derived whether plastic deformation occurs. In that case S_3 is too small related to S_1 to prevent failure of the soil. The minimum magnitude of S_3 to prevent failure due to S_1 can be found from the Mohr-Coulomb equation:

$$S_3 = K_a S_1 - 2c\sqrt{K_a} \tag{10}$$

where:

$K_a = \tan^2 (45° - \phi/2)$
ϕ = angle of internal friction
c = cohesion

For the sandy subsoil of the experimental field $c = 6$ kN.m^{-2} and $\phi = 31°$; c is mainly determined by the negative pressure of the soil moisture in the unsaturated soil.

In addition to the dynamic stresses due to wheel loading there has also to be accounted for the influence of the weight of the soil on the vertical (σ_z) and horizontal (σ_h) stresses. After TSCHEBOTARIOFF (1951) the horizontal static stress is 0.5 times the vertical static stress. In consequence of this the total S_3 increases relatively much more than the total S_1.

If the smallest principle stress S_3, composed of the dynamic part computed by the model and a static part due to the weight of the soil, is smaller than S_3 calculated after eq. (10), plastic deformation will occur. In that case the soil will continue to deform until S_3 is sufficiently increased to counteract the deformation. The compaction due to plastic deformation can be determined from tri-axial tests. The value of S_3 to apply in these tests can be found from eq. (10) by inserting S_1 obtained from the model calculation.

6. COMPARISON OF COMPUTED AND MEASURED SUBSOIL COMPACTION

Using the relationships given in Fig. 8 the increase in bulk density due to elastic deformation can be obtained from the computed mean normal stress (σ_m) (see Fig. 9). The measured increase in bulk density is greater than the compaction due to elastic deformation, what is caused by plastic deformation. The occurrence of plastic de-

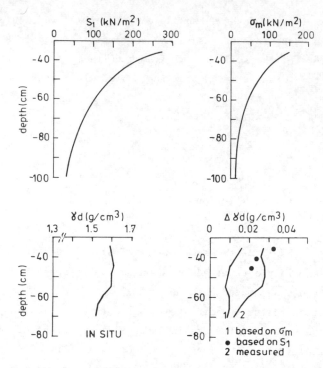

Fig. 9. Computed first principle (S_1) and mean (σ_m) stress distributions with depth, initial dry bulk density (γ_d) in the field
and the increase in bulk density ($\Delta\gamma_d$) based on calculations
with σ_m (1), with S_1 (•) and measured in situ (2)

formations can be concluded from the deformation of the grid shown in
Fig. 5, where decreases in height of the grids are measured up to
5.4%. The plastic deformation necessitates also the incorporation of
the concentration factor ν into the model.

Although much smaller than in the topsoil, the occurrence of
plastic deformation in the subsoil follows also from the computed
principal stresses S_1 and S_3. It appears that S_3 is too small to a
depth of 80 cm to prevent failure of the soil. This agrees with the
measurements in the field, where to a depth of 70 cm a decrease of
height of the grids was observed to such extent that plastic deformation must be occurred.

To determine soil compaction including also the contribution of
plastic deformation tri-axial tests have been carried out whereby S_3

was calculated with eq. (10) from S_1-values computed with the model for depths of 35, 40 and 45 cm respectively (see Section 5).

From these tests it appeared that the S_1-values computed with the model and those obtained from the tests agree very well. The increases in bulk density obtained thus are indicated in Fig. 9 and agree reasonable well with the field measurements. At greater depths the contribution of plastic deformation to compaction decreases. At a depth of 70 cm compaction is caused by the elastic component to a extent of 75%, as appeared when the curve based on σ_m is compared with the measured values.

7. CONCLUSIONS

From computations with the model and in situ measurements it follows that plastic deformations occur to a depth of 70 cm in a sandy soil under a wheel load of 32 kN. This proves also the necessity to incorporate a concentration factor ν into the model.

The presented model enables to determine compaction due to elastic deformation from the computed mean normal stress (σ_m) and relationships between σ_m and change of volume (ε_v) obtained from tri-axial tests. If plastic deformations occur compaction is obtained from tri-axial tests applying a S_3-value based on the greatest principal stress S_1 computed by the model.

To be able to account for plastic deformation in the model calculations, more information is required about the relationship between S_1 and the volume change ε_v.

REFERENCES

BOUSSINESQ, J. 1885. Application des potentiels à l'étude de l'equilibre et du mouvement des solides élastique. Gauthier-Villais, Paris, cited by Fröhlich

FRÖHLICH, O.K. 1934. Druckverteilung im Baugrunde. Verlag von Julius Springer, Wien

KOOLEN, A.J. and H. KUIPERS. 1983. Agricultural Soil Mechanics. Springer-Verlag, Berlin, Heidelberg, New York, Tokyo

SÖHNE, W. 1953. Druckverteilung im Boden und Bodenverformung unter
Schlepperreifen. Grundlagen Landtechnik

TIMOSHENKO, S. and J.N. GOODIER. 1980. Theory of Elasticity, third
edition. Mc Graw-Hill Book Company, New York, London

TSCHEBOTARIOFF, G.P. 1951. Soil mechanics, foundation and earth
structures. Mc Graw-Hill Book Company, New York, London

Méthodes de mesure du retrait vertical d'un sol argileux

V.Hallaire
Institut National de la Recherche Agronomique, Centre de Recherches Agronomiques d'Avignon, Station de Science du Sol, Montfavet, France

RESUME

Le retrait vertical d'un sol argileux est étudié à l'échelle des agrégats (sur photographies), puis à l'échelle de couches élémentaires de sol (au moyen de capteurs de déplacement verticaux).

Au niveau de l'agrégat, les résultats ne montrent pas de différence entre le retrait vertical et le retrait latéral, et l'hypothèse d'isotropie du retrait peut être acceptée. Au niveau de la couche de sol, les résultats varient avec la profondeur : la couche profonde présente un retrait isotrope, tandis que les couches sus-jacentes présentent un retrait anisotrope.

En comparant ce processus à celui de la fissuration, on envisage les conséquences de cette anisotropie du retrait sur la dynamique structurale du sol : création d'une porosité inter-agrégats fine, puis prise en masse.

ABSTRACT

MEASUREMENT METHODS OF THE VERTICAL SHRINKAGE
OF A CLAYEY SOIL

The vertical shrinkage of a clayey soil was studied on aggregates (with photographs) and on soil layers (with vertical displacement transducers).

On the scale of the aggregates, there was no difference between vertical and diametral shrinkage, and it is assumed that the shrinkage was isotropic. On the scale of the soil layers, results varied with depth : the shrinkage was isotropic for the deeper layer, anisotropic for the other ones.

This process is related with the cracking process. The consequences of the anisotropic shrinkage on the development of the soil structure is explained by the opening, and the subsequent closing of thin inter-aggregates voids.

INTRODUCTION

Les sols à argile gonflante possèdent une dynamique structurale provoquée par les variations d'humidité qui causent le retrait ou le gonflement des particules d'argile, entraînant une modification de la densité à l'échelle des agrégats.

Du retrait résulte une augmentation des vides périphériques à ces agrégats, vides qui peuvent s'exprimer sous deux formes :

- les fissures, constituées par les espaces isolant les agrégats ; elles proviennent d'une modification de l'espace poral : une part de la porosité intra-agrégat se transforme en porosité inter-agrégat.

- le tassement, qui provient de l'effondrement gravitaire des agrégats les uns sur les autres, par suite de leur retrait vertical. Le tassement ne peut pas être considéré comme une transformation d'un espace poral en un autre, mais plutôt comme une diminution de l'espace poral total : une part de la porosité intra-agrégat est supprimée du fait de la réduction de l'épaisseur du sol. Pour le distinguer du tassement causé par un agent extérieur, ce tassement résultant du seul retrait vertical sera désigné par le terme d'"affaissement".

L'étude du compactage -dont l'action est essentiellement verticale- et des possibilités de régénérer la structure du sol nécessite une connaissance précise du volume et de la forme des vides produits par le retrait, c'est-à-dire de l'importance relative de la fissuration et de l'affaissement à partir d'un état initial donné.

Cette dynamique porale résultant de l'isotropie ou de l'anisotropie du retrait, il est nécessaire de connaîre les parts respectives du retrait latéral et du retrait vertical. On se propose ici d'étudier, indépendamment de toute intervention mécanique d'origine humaine, l'effet du seul desséchement sur l'affaissement du sol.

LE MATERIAU D'ETUDE

La parcelle étudiée (Les Vignères) est située dans la plaine du Comtat Venaissin, à 20 km à l'Est d'Avignon. Le desséchement in-situ est assuré par une culture de fétuque, implantée depuis 6 ans.

Le sol est développé sur des dépôts alluviaux récents, d'origine fluviale (Rhône-Durance-Calavon) et lacustre (Plaine des Sorgues). Il contient plus de 50 % d'argiles gonflantes sur près de 2 mètres d'épaisseur, sous lesquels on retrouve des lentilles sableuses et des galets.

Tableau 1   Caractéristiques physiques du sol étudié

Profondeur (cote hivernale)	Classes granulométriques (µm) %					Calcaire total %
	0-2	2-20	20-50	50-100	200-2000	
20 cm	51,3	30,4	7,4	8,6	2,3	27,5
40 cm	51,5	29,6	7,3	9,1	2,5	28,2
60 cm	53,2	31,1	6,5	7,3	1,9	29,1
80 cm	54,9	32,6	5,4	5,6	1,5	29,5
100 cm	53,2	34,9	4,5	5,6	1,8	32,3

Dans le domaine des humidités observées au champ, le retrait des agrégats est proche du retrait normal : leur volume massique $1/\gamma_\tau$ est lié linéairement à leur teneur en eau pondérale w , la pente de cette relation est voisine de 1.

Tableau 2   Droites de retrait des agrégats (w > 11 %)

20 cm	$1/\gamma_\tau = 0,881\ w + 0,426$	$r^2 = 0,989$	$n = 25$
40 cm	$1/\gamma_\tau = 0,987\ w + 0,388$	$r^2 = 0,991$	$n = 20$
60 cm	$1/\gamma_\tau = 0,918\ w + 0,399$	$r^2 = 0,992$	$n = 20$
80 cm	$1/\gamma_\tau = 0,898\ w + 0,401$	$r^2 = 0,985$	$n = 20$
100 cm	$1/\gamma_\tau = 0,986\ w + 0,396$	$r^2 = 0,968$	$n = 20$

L'HYPOTHESE D'ISOTROPIE DU RETRAIT

En période hivernale, le sol étant à son niveau de gonflement maximum, la densité apparente sèche du sol en place n'est pas significativement différente de la densité apparente sèche des agrégats ; les espaces inter-agrégats peuvent donc être considérés comme nuls (bien que les fissures ne soient pas décelables, un faible volume fissural, inférieur aux erreurs sur les mesures de densité, peut cependant persister et constituer pour le drainage des voies de passage importantes).

A partir de cet état initial "saturé", le retrait sera dit isotrope, à une échelle donnée, s'il est identique dans les 3 dimensions de l'espace :

$$\frac{\Delta x}{x} = \frac{\Delta y}{y} = \frac{\Delta z}{z}$$

$\frac{\Delta z}{z}$ représente l'affaissement de l'élément de sol considéré, $\frac{\Delta x}{x}$ et $\frac{\Delta y}{y}$ le retrait latéral des agrégats qui le composent.

L'hypothèse d'isotropie ne privilégiant aucune direction, le volume V d'un élément de sol est à tout instant proportionnel au cube de son épaisseur z :

$$\frac{V}{V_o} = \frac{z^3}{z_o^3}$$

où V_o et z_o correspondent à un état initial donné.

Si γ_τ est la densité apparente sèche d'un agrégat de volume V et d'épaisseur z, on a

$$\frac{z^3}{z_o^3} = \frac{V}{V_o} = \frac{\gamma_{\tau o}}{\gamma_\tau}$$

Dans le domaine d'humidité où le volume massique de l'agrégat est proportionnel à son humidité, on a

$$1/\gamma_\tau = a\ w + b$$

L'hypothèse d'isotropie du retrait peut donc s'écrire

$$\frac{z^3}{z_0^3} = \gamma_{\tau_0} \, (a \, w + b)$$

et l'affaissement "isotrope" prend la forme

$$\frac{\Delta z}{z} = 1 - [\gamma_{\tau_0} \, (a \, w + b)]^{1/3}$$

La prédiction de l'affaissement d'un sol argileux gonflant nécessite la vérification de cette hypothèse à deux niveaux :

- au niveau de l'agrégat,

- au niveau d'un ensemble d'agrégats de même constitution et de même humidité, c'est-à-dire au niveau d'une couche élémentaire de sol.

LE RETRAIT VERTICAL DES AGREGATS

Méthode

On a prélevé des agrégats de 5-10 mm en repérant avec précision leur orientation en place. Amenés à saturation, ils ont été déposés sur le plateau d'une balance avec la même orientation, de façon à présenter à l'observateur une face latérale.

Le dessèchement a été fait par évaporation. Tout au long du retrait, les agrégats ont été photographiés et pesés à intervalles réguliers.

A la fin du dessèchement, on a mesuré la masse volumique des agrégats par poussée d'Archimède dans le pétrole (Monnier et al., 1973).

Enfin, sur des agrandissements photographiques, on a déterminé sur plusieurs agrégats leur retrait latéral et leur retrait vertical.

Résultats

La figure 1 présente pour des agrégats prélevés à 40 cm, les retraits latéral et vertical en fonction de l'humidité pondérale. Les points sont donnés avec leur écart-type, calculé sur 6 agrégats. Ces valeurs expérimentales sont comparées à la courbe théorique du retrait isotrope, calculée à partir des mêmes conditions initiales.

Cette figure montre clairement qu'il n'y a pas de différence entre le retrait vertical et le retrait latéral, et que ce retrait peut être considéré comme isotrope à l'échelle de l'agrégat.

Ce résultat confirme les hypothèses formulées par Grossman et al. (1968) et les résultats obtenus par Franzmeier et Ross (1968) sur des mottes naturelles de 5 à 8 cm de diamètre.

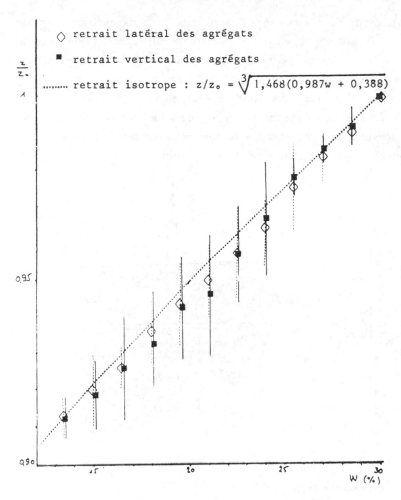

Figure 1. Retrait latéral et vertical des agrégats (40 cm).

LE RETRAIT VERTICAL DES COUCHES DE SOL

Méthode

Quatre couches de sol ont été étudiées : 10-30 cm, 30-50 cm, 50-70 cm et 70-90 cm (ces profondeurs correspondent à des cotes hivernales).

Les variations d'épaisseur des couches de sol ont été mesurées au moyen de deux types de capteurs (figure 2) :

- le capteur de type I, dérivé de celui décrit par Selig et Reinig (1982), est constitué d'un élément fixe ancré en profondeur, et d'un élément mobile à une profondeur déterminée. Les variations lues à la surface du sol correspondent aux variations de cote à cette profondeur ; la variation d'épaisseur d'une couche est donc la différence entre les variations de la cote supérieure et de la cote inférieure de cette couche. Cinq capteurs ont été placés à chacune des profondeurs suivantes : 10 cm, 30 cm, 50 cm, 70 cm et 90 cm.

- le capteur de type II est constitué de deux éléments mobiles, aux cotes supérieure et inférieure de la couche de sol étudiée. Les variations d'épaisseur d'une couche peuvent ainsi être lues directement à la surface du sol. Cinq capteurs ont été placés pour chacune des 4 couches étudiées.

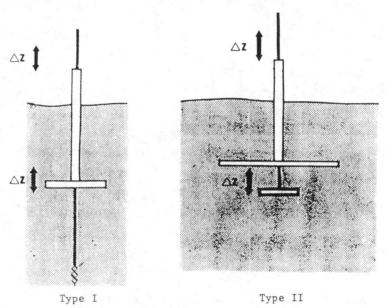

Type I   Type II

Figure 2. Capteurs de retrait vertical.

L'humidité était déterminée après séchage à 105°C d'échantillons prélevés à la tarière au niveau médian de chacune des couches de sol (20, 40, 60 et 80 cm). Pour chaque relevé, 10 profils hydriques étaient effectués. Les valeurs obtenues étaient comparées à celles mesurées par des micropsychromètres (Bruckler, 1984) placés au voisinage immédiat des capteurs de type II.

Le dessèchement du sol s'est déroulé de la fin Mars jusqu'au début Août. Les relevés de cote et d'humidité ont été effectués en moyenne toutes les 48 heures.

Résultats

Les figures 3 et 4 présentent, pour les couches 30-50 cm et 70-90 cm, les variations d'épaisseur en fonction de leur humidité. Les points expérimentaux sont comparés à la courbe calculée selon l'hypothèse d'un retrait isotrope.

Le retrait vertical de la couche 70-90 cm coïncide parfaitement avec la courbe théorique du retrait isotrope. Les agrégats restent donc superposés tout au long du desséchement et leur retrait se traduit entièrement par un effondrement de la couche.

Le retrait vertical de la couche 30-50 cm montre, au contraire, un écart significatif avec le retrait vertical calculé selon l'hypothèse d'isotropie. Pour cette couche (ainsi que pour les couches 10-30 cm et 50-70 cm), la courbe peut être subdivisée en deux phases :

- au début du desséchement, l'affaissement de la couche est inférieur à celui des agrégats. Le retrait vertical a donc pour conséquence la création d'une porosité inter-agrégats ; ceci ne peut s'interpréter qu'en admettant que la gravité ne suffit pas à maintenir les agrégats en contact parfait les uns au-dessus des autres au cours du retrait.

- à partir d'une humidité voisine de 23 %, les points expérimentaux tendent à rejoindre la courbe théorique. Il y a donc fermeture progressive des espaces précédemment ouverts pour atteindre, en fin de desséchement, une superposition parfaite des agrégats.

DISCUSSION

- L'hypothèse d'isotropie, vérifiée au niveau des agrégats, ne se vérifie pas toujours au niveau d'une couche de sol. On assiste même à une inversion de l'anisotropie au cours du desséchement : le retrait vertical est inférieur au retrait isotrope dans certains domaines d'humidité, supérieur dans d'autres domaines. Il n'est donc plus permis de considérer, avec certains auteurs (Reeve et al., 1980 ; Hack, 1984), l'hypothèse d'isotropie inéluctable du fait des seuls effets gravitaires.

- Les vides résultant de cette anisotropie du retrait sont analogues à une "fissuration horizontale". L'existence de cette porosité est une donnée importante, pour la prédiction du compactage : selon le sol et l'humidité où l'on se situe, le rapprochement des agrégats sans déformation est théoriquement possible ou ne l'est pas. Mais il demeure évident que le phénomène est lié également à l'état structural initial : dans le cas étudié de prairie permanente non compactée, on a initialement un volume de fissures nul ou presque, mais des discontinuités mécaniques peuvent persister entre les éléments structuraux ; pour d'autres conditions initiales (compactage à l'état humide),

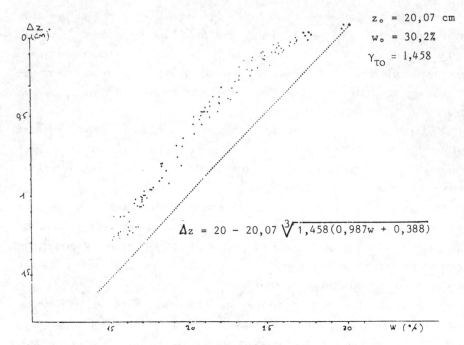

Figure 3. Retrait vertical de la couche 30-50 cm.

$$\Delta z = 20 - 20,07 \sqrt[3]{1,458(0,987w + 0,388)}$$

$z_0 = 20,07$ cm
$w_0 = 30,2\%$
$\gamma_{TO} = 1,458$

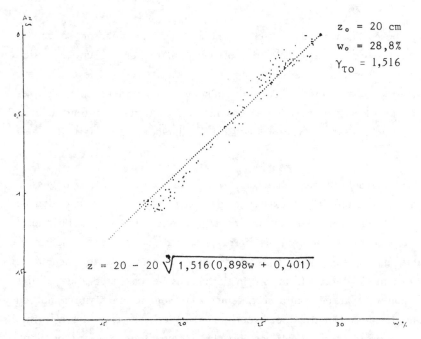

Figure 4. Retrait vertical de la couche 70-90 cm.

$$z = 20 - 20 \sqrt[3]{1,516(0,898w + 0,401)}$$

$z_0 = 20$ cm
$w_0 = 28,8\%$
$\gamma_{TO} = 1,516$

où la structure à saturation serait continue, il est probable que le retrait aurait un effet différent. De ce fait, les relations entre le compactage et la régénération structurale des sols argileux gonflants peuvent jouer même dans le domaine saturé.

- Le processus d'un affaissement en deux temps présente une étroite analogie avec le modèle de fissuration décrit sur le même type de sol (Hallaire, 1984) où l'on observait deux phases de fissuration au cours du retrait :

    - au début du desséchement, le retrait provoquait un réseau homogène et serré de fissures très fines. Cette phase peut être assimilée à la première phase de l'affaissement, où l'on constate la création d'une "fissuration horizontale", suffisamment fine pour que subsistent des points de contact entre les agrégats : ces "ponts" s'opposent à l'effondrement total de la couche de sol.

    - à partir d'une humidité voisine de 23 %, le retrait provoquait une redistribution des fissures, certaines continuant à s'ouvrir largement en "macrofissures", tandis que les autres se refermaient partiellement ou totalement. Cette phase peut être assimilée à la deuxième phase de l'affaissement : à partir de la même humidité, on assiste à une fermeture progressive de la porosité "horizontale", au profit d'un effondrement total du sol.

CONCLUSION

L'analogie entre l'évolution de la fissuration et celle de l'affaissement semble indiquer que le même processus détermine les deux phénomènes ; l'explication de ce processus devrait permettre de préciser la notion de "prise en masse" qui intervient dans la seconde phase du desséchement. Mais dès à présent, la description de ce phénomène en deux temps et la détermination de l'humidité à laquelle on passe d'une phase à une autre, semblent fondamentale pour connaître la dynamique porale de ce type de sol, et pour prévoir les conditions optimales de régénération structurale.

REFERENCES

Bruckler, L., 1984. Utilisation des micropsychromètres pour la mesure du potentiel hydrique du sol en laboratoire et in situ. Agronomie, 4(2), 171-182.

Franzmeier, D.P. and Ross, S.J., 1968. Soil swelling : laboratory measurement and relation to other soil properties. Soil Sci. Soc. Amer. Proc., 32, 573-577.

Grossman, R.B. ; Brasher, B.R. ; Franzmeier, D.P. and Walker, J.L., 1968. Linear extensibility as calculated from natural-clod bulk density measurements. Soil Sci. Soc. Amer. Proc., 32, 570-573.

Hack, H.R.B., 1984. Calculation of the field volumetric water content of cracking clay soils from measurements of gravimetric water content and bulk density. J. Soil Sci., 35, 299-315.

Hallaire, V., 1984. Evolution of crack networks during shrinkage of a clay soil under grass and winter wheat crops. Proc. of ISSS Symposium, ILRI Publication 37, Wageningen.

Monnier, G. ; Stengel, P. et Fies, J.C., 1973. Une méthode de mesure de la densité apparente des petits agglomérats terreux. Application à l'analyse des systèmes de porosité du sol. Ann. agron., 24(5), 533-545.

Reeve, M.J. ; Hall, D.G.M. and Bullock, P., 1980. The effect of soil composition and environmental factors on the shrinkage of some clayey british soils. J. Soil Sci., 31, 429-442.

Selig, E.T. and Reinig, I.G., 1982. Vertical soil extensometer. Geotechn. Test. J., 5(3/4), 76-84.

Fissuration d'un sol argileux gonflant après compactage: Effet de l'humectation

P.Stengel & M.Bourlet
Institut National de la Recherche Agronomique, Centre de Recherche Agronomique d'Avignon, Station de Science du Sol, Montfavet, France

RESUME

L'effet de l'humectation sur la fissuration dans un sol argileux initialement compact a été étudié sur des échantillons remaniés en laboratoire, et in-situ.

Au laboratoire, des massifs cylindriques à structure continue ont été amenés à différentes teneurs en eau, puis réhumectés par mise en contact avec une couche de sable saturée pendant 72h. Le potentiel de l'eau dans la couche de sable était maintenu constant. Deux séries d'expériences ont été réalisées, ce potentiel étant fixé respectivement à 0 et -10 mbar. Les fissures ne sont apparues que dans les échantillons dont l'humidité initiale était inférieure à celle du point d'entrée d'air et mis en contact avec l'eau à potentiel nul. La longueur de fissures formées était une fonction décroissante de l'humidité initiale et s'annulait à l'humidité du point d'entrée d'air.

Au champ, différentes doses d'irrigation ont été apportées sur un sol initialement sec, préalablement compacté par roulage. Les doses d'apport ont été de 3,7; 7,5; 15 et 30 mm. Après irrigation, la surface irriguée a été recouverte d'une feuille de plastique pour éviter l'évaporation. La fissuration à la surface du sol a été décrite avant et 24h après l'arrosage. L'accroissement de longueur de fissure consécutif à l'irrigation était maximal pour la dose la plus faible et nul pour la dose de 30 mm. Du fait de la redistribution de l'eau apportée après la fin de l'irrigation, il est impossible de conclure quant à l'effet direct de l'humectation sur le développement des fissures dans ces conditions d'observation.

Les résultats montrent que l'humectation peut jouer un rôle direct et important dans la régénération de la structure d'un sol argileux gonflant après compactage, surtout à proximité de la surface. Les conditions dans lesquelles la fissuration au gonflement se manifeste in-situ doivent être précisées expérimentalement et par la mise au point du modèle tenant compte de la variation des propriétés mécaniques du sol avec l'humidité.

ABSTRACT

EFFECT OF WETTING ON CRACKS FORMATION IN A COMPACTED CLAY SOIL

Effects of wetting on cracks development in a compacted clay soil were studied both in laboratory and field experiments.

In the laboratory remoulded samples were prepared in order to obtain initially continuous (crackless) and compact soil cores, with different water content. The cores were rewetted by putting it in contact with saturated sand during 72h. Water potential in the sand was maintained at a constant value. Two values were compared : 0 and -10 mbar.

Cracks developped during wetting only when initially unsaturated cores were rewetted with free water. Length of cracks which were formed during wetting increased as initial water content decreased under air-entry value.

In the field, water was spraied on initially compacted and superficially dry soil. Different irrigation dosis were supplied, varying from 3.7 to 30 mm. After irrigation, the soil surface was covered with a plastic sheet to avoid evaporation. Increase in cracks-length was measured by image-analysis of soil surface photographies before and 24h after irrigation. Crack-length increase was maximum for the lowest irrigation dosis, and decreased to zero with the 30 mm dosis. Due to water movements after irrigation, it cannot be concluded that cracks formation occured during wetting.

Results show that cracking process during wetting can be very important for soil structure regeneration in compacted clay soils, mainly near soil surface. Division of compacted structural units can be very intense in only one wetting phase. Quantitative assessment of its effects in the field requires further experiments in field or laboratory conditions, prior to intends of modelling.

INTRODUCTION

La fissuration par retrait-gonflement constitue avec le travail du sol et le gel, un des processus principaux de formation d'une structure fragmentaire. Son rôle dans la régénération de la structure après compactage peut être essentiel. C'est le cas dans les horizons non travaillés, près de la surface lorsqu'on pratique le semis direct, ou après un compactage profond. C'est également le cas pour la fragmentation des mottes compactes des couches travaillées qui peut être très couteuse, voire impossible, par recours à des moyens mécaniques.

Si l'importance du phénomène est reconnue, et dès longtemps utilisée par les agriculteurs travaillant des sols argileux, la connaissance des processus de fissuration reste très incomplète. Les modèles mécaniques de fissuration (HETIARATCHI et O'CALLAGHAN, 1980) et les données concernant son déterminisme physique sont insuffisants pour prévoir la vitesse de développement des fissures et l'évolution de leur morphologie, c'est-à-dire l'intensité des processus de régénération structurale après compactage. Les descriptions précises elles-mêmes sont rares, du fait de la lourdeur des observations nécessaires et des difficultés techniques rencontrées (RINGROSE-VOASE et BULLOCK, 1984). De ce fait, les conditions de formation des fissures sont mal définies en dehors d'une simple analyse du bilan en volume du retrait-gonflement. Celle-ci à un caractère très réducteur, car le volume fissural n'a que peu de relations avec les conséquences de la fissuration sur les propriétés mécaniques et de transfert du sol.

Elle est, de plus, fréquemment erronée car fondée sur l'hypothèse d'isotropie des déformations qui n'est pas généralement vérifiée (HALLAIRE,

1985). L'analyse mécanique prévoit que le retrait et le gonflement sont chacun susceptibles de causer la rupture du matériau sol. Elle distingue (RAATS, 1984) :

- l'apparition de fissures dues aux forces de traction
- le développement de surfaces de glissement résultant du cisaillement.

La distinction morphologique entre les pores planaires engendrés par ces deux types de rupture repose en premier lieu sur leur orientation par rapport aux directions des contraintes principales. Dans ce qui suit, nous n'avons pas tenté d'établir cette classification et nous appellerons fissure indiféremment tout pore planaire formé par retrait ou gonflement.

Le rôle du gonflement dans la fissuration est surtout invoqué pour expliquer l'existence des slicken sides et la forme des éléments structuraux dans des horizons profonds (YONG et WARKENTIN, 1975 ; BLOKHUIS, 1982). Les hypothèses concernant les conditions nécessaires incluent le confinement latéral (fermeture préalable des fissures formées par dessiccation) et vertical (poids des couches sus-jacentes). En revanche, il existe peu d'informations sur le rôle spécifique de l'humectation dans la fissuration à proximité de la surface du sol. Il s'agit ici de connaître les effets de phase de gonflement beaucoup plus limitées dans le temps, consécutives à un épisode pluvieux ou une irrigation, et en conditions non confinées puisque la pression pédostatique est très faible et que le gonflement n'aboutit pas nécessairement à obturer le volume de pores structuraux préexistant.

Les travaux expérimentaux présentés dans cet article avaient pour but d'estimer l'importance possible de ces effets dans la régénération d'une structure fragmentaire après compactage des couches superficielles.

MATERIELS ET METHODES
1. Le Sol

Les travaux expérimentaux de terrain et de laboratoire ont porté sur la couche de surface d'un sol argileux dont les caractéristiques analytiques figurent au tableau 1. Ses propriétés de gonflement-retrait sont décrites par la courbe de la figure 1. Elle a été obtenue par mesure de la densité d'agrégats naturels, séparés par tamisage entre 2 et 3 mm (PINOCHET, 1981 ; STENGEL, 1982). On constate que l'amplitude totale du retrait est importante et qu'à l'échelle des agrégats ce retrait est assimilable au retrait normal jusqu'à des valeurs très basses du potentiel de l'eau.

97

Tableau 1 : Caractéristiques analytiques du sol étudié.

Analyse granulométrique (%)					Matière organique (%)	CaCO$_3$ %	CEC meq/100 g	pH
<2 µm	2-20 µm	20-50 µm	50-200 µm	200-2000 µm				
48,0	32,8	8,5	9,0	1,7	2,42	26,9	20,1	8,2

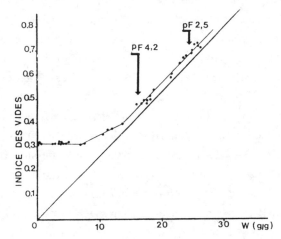

Figure 1 : Variations de l'indice des vides d'agrégats en fonction de leur teneur en eau massique.

2. Expériences de laboratoire

2.1. Préparation des échantillons.

Le matériau prélevé dans la couche de sol précédemment décrite a été séché à l'air, puis forcé à travers un tamis de 2 mm. Il a ensuite été amené à une humidité correspondant à sa limite de liquidité par agitation mécanique d'une durée de 20 minutes en présence de la quantité d'eau nécessaire. L'agitation a été répétée deux fois à 24h d'intervalle. Il a subi un ressuyage de deux semaines sur une couche du même matériau sec et tamisé. Le massif cylindrique obtenu a alors été découpé en trois tranches. Les tranches extrêmes ont été utilisées pour les mesures de teneur en eau et densité. La partie médiane, d'une hauteur d'environ 4 cm, a été desséchée dans l'atmosphère du laboratoire jusqu'à la teneur en eau recherchée. Seules les deux sections planes des cylindres étaient en contact avec l'air.

Ce mode de préparation a permis d'obtenir des massifs ne présentant aucune fissure décelable, ni visuellement, ni mécaniquement, ni par mesure de la densité.

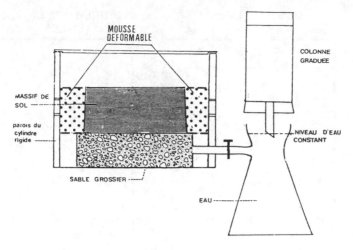

Figure 2 : Schéma du dispositif de réhumectation.

2.2. Réhumectation

Les massifs cylindriques ont été posés sur une couche de sable grossier saturé (figure 2), alimentée en eau par un vase de Mariotte permettant simultanément de contrôler le niveau de l'eau apportée et de mesurer le volume d'eau absorbé par le sol. Dans les expériences dont nous présenterons en détail les résultats, le niveau de l'eau affleurait à la base du massif en réhumectation qui était donc en contact avec de l'eau à potentiel nul. L'ensemble du massif était enfermé dans une enceinte cylindrique étanche ayant pour but d'empêcher l'évaporation. La durée de réhumectation a été limitée arbitrairement à 72h.

2.3. Mesures

Au cours du temps, les variations de hauteur et de diamètre des massifs réhumectés ont été mesurées, ainsi que les quantités d'eau absorbée. Simultanément, la photographie de la section supérieure du massif a permis de suivre l'évolution de la fissuration. Les photographies ont ensuite été numérisées et les images analysées sur un analyseur d'image PERICOLOR 2000, de NUMELEC[*], de façon à mesurer :

[*]   La marque et le type de l'appareil ne sont mentionnés ici que pour l'information du lecteur. Ceci n'implique aucune caution ou garantie de la part des auteurs, ou de celle de l'INRA en faveur du matériel nommé, pas plus que d'intention commerciale.

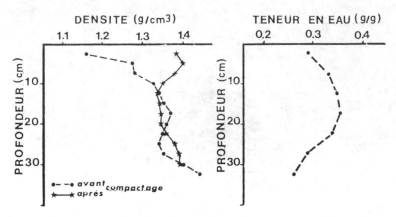

Figure 3 : Profils de teneur en eau et densité avant et après compactage.

- la porosité de fissure

- la longueur totale des fissures et la distribution des longueurs par classes de largeur, que nous appellerons par la suite granulométrie des fissures.

3. Expériences de terrain

3.1. Préparation du sol.

Le sol labouré en automne, puis travaillé superficiellement au printemps, et laissé nu, a reçu une irrigation de 40 mm de façon à le ramener, sur toute l'épaisseur de la couche travaillée, à une humidité voisine de la capacité de rétention. Après un ressuyage de 24h, il a été compacté par passage répété d'un rouleau. Les profils de teneur en eau et de densité obtenus avant et après compactage sont présentés à la figure 3.

Après compactage, le sol protégé de la pluie par un tunnel plastique, s'est desséché par évaporation pendant une dizaine de jours. Au cours de ce desséchement, un réseau de fissures s'est développé.

3.2. Réhumectation.

L'arrosage a été réalisé manuellement par aspersion, à l'aide d'un pulvérisateur de jardin. Le débit d'apport étant constant (15 mm/h), les différentes doses d'apport correspondent à des durées d'irrigation différentes. Les doses réalisées ont été de : 3,5; 7,5; 15 et 30 mm. La surface arrosée comprenait dans chaque cas deux carrés de 50 x 50 cm², l'un étant réservé aux mesures, l'autre aux observations en surface des fissures et à la prise de vue photographique. Après arrosage, elle était recouverte d'une feuille de plastique de façon à éviter l'évaporation.

3.3. Mesures.

Les mesures ont été pratiquées pour chaque traitement immédiatement avant et 24h après le début de l'irrigation. Elles ont comporté la mesure de l'humidité et de la densité entre 0 et 35 cm de profondeur ainsi que celle du gonflement vertical.

La morphologie des fissures à la surface du sol a été étudiée par analyse des images photographiques. Les variables morphologiques mesurées ont été les mêmes que pour les échantillons de laboratoire.

III. RESULTATS

1. Echantillons remaniés en laboratoire

1.1. Amplitude du gonflement.

Les caractéristiques initiales et finales des différents massifs soumis à réhumectation sont présentées au tableau 2. On constate que, quelle que soit l'humidité (ou le potentiel de l'eau) initiale, tous les massifs ont subi un gonflement. L'amplitude de ce gonflement décroît quand l'humidité initiale augmente :

- d'une part, parce que l'écart d'humidité par rapport à l'équilibre à potentiel $\simeq 0$ est plus faible,
- d'autre part, parce que les massifs initialement les plus humides ont des teneurs en eau plus faibles après 72H de réhumectation.

Cette dernière constatation implique que ces massifs n'ont pas atteint en 72h la teneur en eau d'équilibre. Nous aurons à revenir sur cette différence entre les cinétiques d'humectation et ses relations avec la fissuration observée.

Tableau 2 : Caractéristiques physiques des massifs réhumectés.

	Domaine de retrait	Non Retrait			Retrait résiduel				Retrait Normal	
ETAT INITIAL	Humidité g/g	0,040	0,056	0,071	0,105	0,125	0,143	0,153	0,177	0,207
	Densité sèche g/cm³	1,93	1,89	1,92	1,92	1,86	1,80	1,77	1,78	1,66
	Potentiel bar	-900	-660	-470	-220	-140	-96	-78	-54	-24
ETAT FINAL	Humidité g/g	0,364	0,397	0,330	0,340	0,301	0,307	0,238	0,250	0,256
	Densité g/cm³	1,25	1,19	1,34	1,34	1,43	1,35	1,57	1,62	1,58

Figure 4 : Photographies de la face supérieure des cylindres en fin de réhumectation.

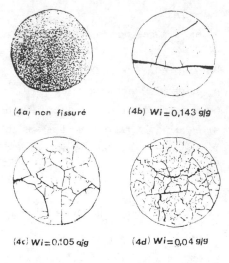

(4a) non fissuré

(4b) Wi = 0,143 g/g

(4c) Wi = 0,105 g/g

(4d) Wi = 0,04 g/g

1.2. Fissuration en fin d'humectation.

La courbe de retrait des massifs en desséchement permet de situer les différentes humidités initiales par rapport au point d'entrée d'air et à la limite de retrait. Le rapprochement avec les photographies de la figure 4 montre ainsi que :

- les échantillons initialement saturés n'ont pas été fissurés au cours de l'humectations (4a)
- dans tous les échantillons initialement désaturés des fissures sont apparues (4b, 4c, 4d)
- l'intensité de la fissuration croît quand l'humidité initiale décroît en dessous de l'humidité à l'entrée d'air.

La possibilité d'une contribution importante du phénomène de fissuration par humectation à l'évolution structurale des couches de surface apparaît donc clairement. Les photographies des échantillons initialement les plus secs (humidité inférieure à celle de la limite de retrait) montrent qu'il peut aboutir à la formation d'une structure fine en un seul épisode d'humectation.

L'effet de l'humidité initiale sur la fissuration est exprimé en termes quantitatifs par la figure 5. La longueur de fissure par unité de surface décroît, de façon apparemment linéaire, avec l'humidité, jusqu'au point d'entrée d'air où elle s'annule. Aucune singularité n'est décelable à l'humidité de la limite de retrait.

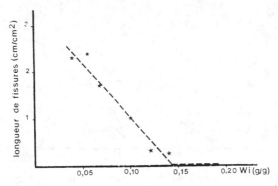

Figure 5 : Relation entre la longueur des fissures par unité de surface L_F et l'humidité initiale du massif.

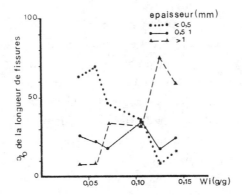

Figure 6 : Influence de l'humidité initiale sur la granulométrie des fissures.

La granulométrie des fissures varie également en fonction de l'humidité initiale (figure 6). La proportion des fissures les plus fines (e ⩽ 0,5 mm), qui représentent environ 70 % de la longueur totale dans les échantillons les plus secs, décroît avec l'humidité et atteint seulement 10 % pour des humidités initiales voisines du point d'entrée d'air. Cette diminution est compensée par un accroissement équivalent de la proportion des fissures les plus larges (e > 1 mm), tandis que la classe intermédiaire représente une fraction peu variable du total.

Les échantillons initialement les plus secs ont donc, après humectation, une structure plus finement fragmentée, correspondant à une longueur de fissures plus grande, et des pores fissuraux de plus faible épaisseur. La réduction d'épaisseur est certainement liée au fait que le développement des fissures fines a lieu, au moins en partie, au détriment du volume des fissures larges. La relation entre les volumes respectifs des différentes classes

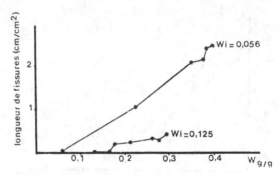

Figure 7 : Variation de la longueur des fissures visibles au cours de l'humectation.

dépend donc des conditions de frettage des massifs. Dans le cas présent, ce frettage était assuré par une couche de mousse qui a nécessairement limité le gonflement malgré sa grande souplesse, au moins dans la phase correspondant aux potentiels de l'eau les plus élevés.

1.3. Evolution de la fissuration au cours de l'humectation.

Les variations de la longueur de fissure par unité de surface en fonction de l'humidité moyenne sont présentées (figure 7) pour deux massifs d'humidités initiales différentes. On constate que l'écart entre les longueurs finales peut difficilement être imputé à la différence d'humidité en fin de réhumectation.

Il existe pour des humidités moyennes identiques dans les deux massifs et tend à croître au cours de l'humectation.

Bien que cet aspect ne puisse être développé dans le cadre de cet article, il convient, par ailleurs, de préciser que les différences d'humidité entre massifs ne correspondent que partiellement à des différences de potentiel de l'eau en fin d'humectation. Elles sont dues pour une large part à la quantité d'eau retenue dans les fissures elles-mêmes, dont le taux de saturation est de l'ordre de 60 %.

L'observation du phénomène de fissuration a montré que les premières fissures apparaissent très rapidement à la face supérieure des massifs les plus secs, bien avant que le front d'humectation par diffusion en phase liquide ait pu l'atteindre. Les fissures formées par rupture de la partie non humectée du massif jouent alors le rôle de chemins préférentiels dans lesquels se produit un transfert rapide de type capillaire. A partir des parois de ces chemins préférentiels se produit une diffusion dans les volumes non fissurés et secs. La photo de la figure 8 illustre clairement ce phénomène dont le

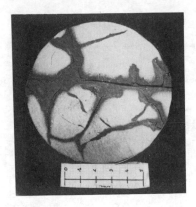

Figure 8 : Photographie de la face supérieure d'un cylindre en cours de réhumectation (Teneur en eau initiale : 7,1 %).

rôle dans la cinétique d'humectation des massifs est essentiel. Les gradients intenses de potentiel et de déformation qui en résultent provoquent ensuite une fissuration plus fine.

1.4. Influence du potentiel de l'eau apportée.

Des massifs préparés suivant le même protocole, et amené à des humidités couvrant le même intervalle de variation, ont été soumis à réhumectation par mise au contact avec de l'eau soumise à une succion de 10 mbar.

Aucun d'entre eux n'a fissuré. Il semblerait donc que la mise en contact avec de l'eau à un potentiel très voisin de zéro soit une condition nécessaire à la fissuration par humectation.

2. Fissuration in-situ

2.1. Variation d'humidité

Les profils hydriques avant irrigation et au moment de l'observation des fissures sont représentés à la figure 9, pour les différentes doses d'irrigation, à l'exception de la plus faible (3,7 mm) pour laquelle l'échantillonnage dans la zone humectée était très difficile.

Dans tous les cas, l'humidité initiale à proximité de la surface était telle que les mottes comprises entre les fissures étaient désaturées et même à une humidité inférieure à la limite de retrait.

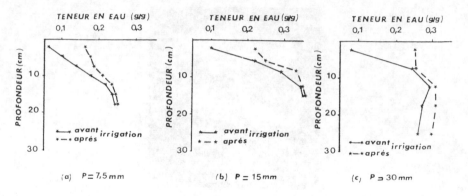

Figure 9 : Profils hydriques avant et après irrigation.

Au moment de l'observation, pour les doses de 7,5 et 15 mm, les teneurs en eau étaient croissantes avec la profondeur dans la zone humectée. En condition d'évaporation très réduite, ceci ne peut s'expliquer qu'en supposant qu'une partie de l'eau apportée s'est infiltrée directement dans les fissures préexistantes, formées par dessiccation. Il est certain d'autre part, que pendant le temps séparant la fin de l'irrigation de l'observation, une redistribution s'est opérée entre la surface ayant reçu directement l'irrigation et les couches immédiatement sous jacentes. En effet, l'infiltrabilité des mottes tassées (non fissurées) étant très faible, de l'eau libre était présente à leur surface pendant l'irrigation. Cette surface a donc été portée temporairement à potentiel nul, puis a perdu de l'eau après arrêt de l'irrigation. Ceci n'est pas décelable sur les profils de teneur en eau du fait de l'épaisseur trop grande des prélèvements qui ne permet pas d'estimer le gradient à proximité de la surface.

2.2. Influence de la dose d'irrigation sur la fissuration

A titre d'exemple, les photos de la figure 10 illustrent l'effet d'une irrigation de 7,5 mm sur la fissuration. La figure 11 permet de quantifier le développement très apparent des fissures. L'accroissement de la longueur de fissure par unité de surface ΔL_F est maximal pour la dose d'irrigation P la plus faible (3,7 mm) et nul pour la dose la plus élevée (30 mm).

La fonction $\Delta L = F(P)$ étant nécessairement nulle pour P = o, ces résultats soulèvent la question de la forme et de la continuité de cette fonction au voisinage de l'origine, et en particulier celle de l'existence possible d'un seuil P_o nécessaire au développement de la fissuration.

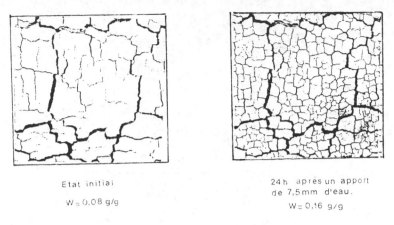

Etat initial

W = 0,08 g/g

24 h après un apport
de 7,5mm d'eau.

W = 0,16 g/g

Figure 10 : Photographie de la surface du sol avant et après irrigation
(P = 7,5 mm).

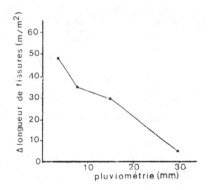

Figure 11 : Influence de la dose d'irrigation sur l'accroissement de la longueur
de fissures L_F.

Pour la dose la plus élevée, l'absence de développement de fissures visibles peut être interprétée de deux façons différentes :

- ou bien la fissuration observée ne s'est produite dans aucun cas durant la phase d'humectation mais pendant la phase de redistribution de l'eau qui l'a suivie. Elle n'est pas apparue après l'irrigation de 30 mm parce que l'humidité de la couche superficielle s'est maintenue à une valeur nettement supérieure à celle des autres doses jusqu'au moment de l'observation (figure 10).

- ou bien la fissuration due à l'humectation n'est pas visible en surface lorsque l'humidité est élevée, car l'épaisseur des fissures est très réduite par le gonflement. Les fissures peuvent également avoir été obturées partiellement par dégradation de la structure de surface sous l'effet de

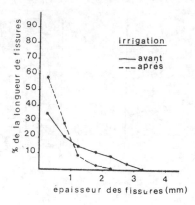

Figure 12 : Variation de la granulométrie des fissures après irrigation (P = 7,5 mm).

l'impact des gouttes, et particulièrement pour la dose d'irrigation la plus forte.

L'accroissement de longueur des fissures n'affecte que les classes de fissures les plus fines (figure 12), principalement celles dont l'épaisseur est inférieures à 0,5 mm. Au contraire, la longueur des fissures les plus larges (**e** > 1,5 mm) décroît après irrigation. Deux phénomènes concourent à cette variation :

      - d'une part, les fissures formées après arrosage sont plus fines que celles qui étaient apparues au cours du retrait,

      - d'autre part, le gonflement a réduit l'épaisseur des fissures prééxistantes.

IV. DISCUSSION

Les résultats du laboratoire établissent l'existence d'un processus de fissuration à l'humectation susceptible d'intervenir de façon importante dans l'évolution structurale à proximité de la surface du sol. Ils permettent de formuler des hypothèses quant à ses conditions d'apparition, qui sont les suivantes :

      - non saturation des éléments structuraux qui subissent l'humectation

      - valeur très voisine de zéro du potentiel de l'eau à la limite de ces
         éléments structuraux.

D'autre part, ils montrent l'existence d'une relation étroite entre l'humidité avant réhumectation et l'intensité de la fissuration.

La généralisation de ces résultats et l'étude des limites de validité des hypothèses suppose la mise au point d'un modèle de rupture, incluant la modélisation des transferts d'eau et des variations de contrainte à l'intérieur du volume de sol humecté. Cette modélisation déjà complexe pour des volumes continus dans le cas des sols argileux, devient extrêmement difficile dès la formation des premières fissures, du fait des hétérogénéités considérables qu'elles induisent.

A partir de ces hypothèses, on peut proposer d'évaluer la probabilité d'apparition d'une fissuration à l'humectation à partir :

      - d'un modèle de variation de l'humidité des couches de surface de sol déterminant la probabilité de désaturation des éléments structuraux non fissurés,

      - de la connaissance de l'infiltrabilité de ces éléments structuraux et de l'intensité des précipitations déterminant la probabilité d'obtention d'un potentiel nul en surface.

Dans le cas de l'expérimentation de terrain, les conditions de fissuration à l'humectation définies précédemment étaient réunies, et l'accroissement constaté de la longueur de fissures apparaît en bon accord avec les résultats de laboratoire. Cependant, les conditions d'observation choisies ne permettent pas de conclure de façon rigoureuse quant à la nature du processus de formation des fissures développées à la surface. Elles peuvent s'être formées pendant l'humectation ou pendant la phase de décroissance de l'humidité de la couche de surface qui a nécessairement suivi l'irrigation (diffusion de l'eau vers les couches intérieures plus sèches).

L'effet direct de l'humectation doit donc faire l'objet d'études plus adaptées à sa mise en évidence. Celles-ci pourraient être réalisées :

      - en observant le développement des fissures dans la zone sous jacente à la couche saturée par l'apport d'eau, et ceci dès la fin de l'irrigation.

      - en se plaçant dans des conditions où le gonflement ne risque pas d'entraîner la fermeture des fissures, par exemple en étudiant la fissuration de mottes compactes séparées par un grand volume poral.

V. CONCLUSION

La fissuration due à l'humectation peut jouer un rôle spécifique et important dans l'évolution structurale des sols gonflants après compactage. En effet, elle peut aboutir, dans certaines conditions, à une fragmentation fine de volumes initialement continus et compacts, et ceci en une seule phase d'humectation. Il est donc nécessaire de définir avec précision ces conditions pour évaluer la probabilité de leur réalisation sur le terrain.

Les premiers résultats suggèrent que les effets de ce processus sont surtout intenses à proximité immédiate de la surface du sol ou peuvent être réunies les conditions d'état initial très sec et de mise en contact avec de l'eau libre. On peut également supposer qu'ils se manifestent en profondeur à la périphérie d'éléments structuraux lorsque de l'eau libre circule sur les parois de pores grossiers. Ce peut être le cas, par exemple, dans les fissures de grandes dimensions se formant sous culture en période sèche.

REFERENCES BIBLIOGRAPHIQUES

BLOKHUIS W.A., 1982 - Morphology and genesis of vertisols. Transactions of the 12th International Congress of Soil Science. New Delhi (India), 3, 23-47.

HALLAIRE V., 1986 - Méthodes de mesure du retrait vertical d'un sol argileux. Atelier CCE, Avignon (France), 17-18 Septembre 1985, This issue.

HETTIARATCHI D.R.P., 0'CALLAGHAN J.R., 1980 - Mechanical behaviour of agricultural soils. Journal of Agricultural Engineering Research, 25, 239-259.

PINOCHET X., 1981 - Prévision du retrait-gonflement d'agrégats de sol. Relation avec le potentiel de l'eau. Mémoire ENITA. Station de Science du Sol d'Avignon, 35p.

RAATS P.A.C., 1984 - Mechanics of cracking soils. Proceedings of the I.S.S.S. Symposium on Water and Solute Movement in heavy clay soils. Wageningen (The Netherlands), 23, 38.

RINGROSE-VOASE A.J., BULLOCK P., 1984 - The automatic recognition and measurement of soil pore types by image analysis and computer programs. Journal of Soil Science, 35, 673-684.

STENGEL P., 1982 - Swelling potential of soil as a criterium of permanent direct-drilling suitability. 9th Conference of I.S.T.R.O., Osijek (Yugoslavia), 131-136.

YONG R.N., WARKENTIN B.P., 1975 - Soil properties and behaviour. Elsevier Publ. Co., Amsterdam, 449p.

Pore size contribution to the drying process of soil with low moisture content

A.D.Louisakis
Land Reclamation Institute, Sindos, Greece

ABSTRACT

Water transfer, in vapour phase, was measured using three soil fractions at the same bulk density. The data were obtained from a transient desorption experiment conducted at 20°C. Analysis of water content distribution curves showed that soil aggregation had no effect on desorption flux and drying depth. Calculated water diffusivities as a function of water content suggest that tortuous factor does not influence water vapour movement in dry range of soils.

INTRODUCTION

Most studies of water desorption in soils assume that tortuosity plays an important role in soil water flux. On the other hand, soil water distribution, as a function of distance, has been reported to depend on the initial wetness (Keulen and Hillel, 1974; Manenti, 1984; Jackson, 1964). Thus, the resulting profile shape of water content vs. distance is related to the diffusivity for liquid and vapour phase of water flow. The moisture dependent diffusivity has a peculiar "hooked" shape as the soil approaches dryness due to the combination of liquid, vapour and film flows (Gardner, 1959; Hanks and Gardner, 1965; Porter et al. 1960).

The aim of the present paper is to identify those factors that should be carefully taken into account to permit reasonable explanation of this phenomenon.

MATERIALS AND METHODS

The soil used was a loam that had been ground and passed through 2.0, 1.0 and 0.5 mm sieves. Thus three soil fractions were obtained and moistened to the same initial water content, ($\theta i=9.3\%$ d.w.), using microwave technique (Horton et al. 1982). Soil columns 35 cm long and 4.897 cm inside diameter, composed of 10, 10, and 9 glass rings, 5 mm, 10 mm and 20 mm thick respectively. The rings were held together with water-proof tape to make tubes. The columns were packed with the previously mixed soil samples and the thinest rings at the top of each tube.

All soil columns, with closed ends, were scanned using the γ-ray

111

apparatus and the wet bulk densities measured. From the set of the fifteen scanned soil columns, only six were selected, two from each soil fraction. The bulk density was 1,17±0.007 and 1,17±0.009 cm/cm³ among soil columns and along them, respectively.

A ring of 5 cm thick was packed with relatively wetted sodium chloride for each soil column, and then they were placed into an oven (105°C). The oven dry rings were cooled of in the decicator. Then one ring was fitted to the top of each soil column and held with waterproof tape, keeping an air gap of 3 mm between salt and soil, by a glass ring. The columns were immediately closed at the ends and placed almost horizontally (lightly inclined to the salt end) in constant temperature room at 20±0.1°C.

After a certain time the columns were quickly sectioned and the water content of the soil in each ring determined gravimetrically. This operation required 5 to 8 minutes. Plots of water content versus $x.t^{-1/2}$ from the top of soil columns were made and the diffusivity function was calculated.

It must be noted that three assumptions were made in this experiment. They are:

   a) For t>0, the water content at x=0 was maintained at constant value
   b) The flux was one dimension and
   c) No temperature gradient along the columns was developed and the pneumatic pressure was unchanged.

RESULTS AND DISCUSSION

Water content distribution curves plotted as θ versus $\lambda = x.t^{-1/2}$ for each diffusion time period (Fig. 1 and 2). Each set consists of data from three columns with different soil fraction, subjected to the same time

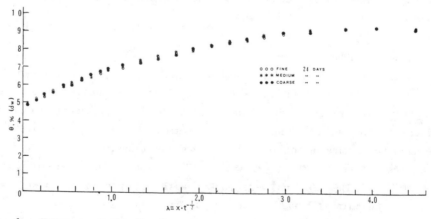

Fig. 1 - Water content, θ, distributions versus $x.t^{-1/2}$

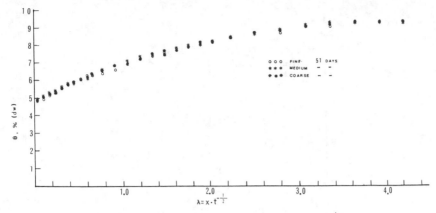

Fig. 2 - Water content, θ, distributions versus $x.t^{-1/2}$

period. The diffusion time period for each run are noted in the figures, and the data points fall close to a single curve. Hence, some of the previously listed assumptions were met and for these conditions, diffusion theory will adequately describe the desorption process. As it can be seen the initial and boundary conditions were the same for all runs.

Applying Bruce and Klute (1959) method, the water diffusivities were calculated for each set (Fig. 3). Water diffusivity points define the hooked portion of the diffusion coefficient versus water content.

From this figure it is evident that the pore-size distribution has no effect on the water diffusivity in dry range of soil. Therefore, the tortuosity can be ignored in soil water transport studies, when the water movement is occurred in vapour phase.

Water content distribution curves (Fig. 1 and 3) indicate that the gravimetric water content of the soil in vapour equilibrium with saturated sodium chloride solution was ~4.7 per cent. Also the effectiveness of the used tape in sealing of the tubes is indicated by the fact that the water contents at the plateaus were the same as the initial water content.

From the study of the above figures one is led to the conclusion that the drying depth was the same for different soil pore size distributions, while all columns had the same total porosity.

SUMMARY

A three set experiment was conducted with two runs for the study of pore-size contribution to the soil drying process. The experiment was conducted with presectioned glass soil columns, using three size fractions of the same soil. Soil samples were moistened at the same water content, using microwave techniques. All soil columns were scanned by the γ-radiation apparatus in order to avoid discrepancies in soil bulk density

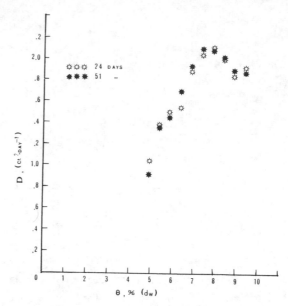

Fig. 3 - Soil water diffusivity, D, versus soil moisture content, θ.

along the columns and between them. The experiment was conducted under constant temperature conditions.

Analysis of collected data confirm the notion that the soil aggregation has no effect on the desorption flux and its depth. It can be mentioned that the diffusion coefficient must not be treated as a constant, even in a narrow soil moisture range when water movement, in porous media, is occurred in vapour phase.

REFERENCES

Bruce, R.R., and Klute, A. 1959. The measure of the soil moisture diffusivity. Soil Sci. Soc. Am. Proc. 20, 458-462.

Gardner, W.R., 1959. Solutions of the flow equation for the drying of soils and other porous media. Soil Sci. Soc. Am. Proc. 23, 183-187.

Hanks, R.J., and Gardner, R.H. 1965. Influence of different diffusivity-water content relations on evaporation of water from soils. Soil Sci. Soc. Am. Proc. 29, 495-498.

Horton, R., Wierenga, J.P., and Nielsen, R.D. 1982. A rapid technique for obtaining uniform water content distributions in unsaturated soil columns. Soil Sci. 133, 397-399.

Jackson, D.R. 1964. Water vapor diffusion in relatively dry soil. II Desorption experiments. Soil Sci. Soc. Am. Proc. 28, 464-466.

Keulen, H. Van. and Hillel, D. 1974. A simulation study of the drying-front phenomenon. Soil Sci. 118, 270-230.

Manenti, M. 1984. Physical aspects and determination of evaporation in deserts. I.C.W. Wageningen. pp. 27-38.

Porter, L.K., Kemper, W.D., Jackson, D.R., and Steward, B.A., 1960. Chloride diffusion in soils as influenced by moisture content. Soil Sci. Soc. Am. Proc. 24, 460-463.

Soil compaction and tillage practices in Ireland

R.A.Fortune
The Agricultural Institute, Oak Park Research Centre, Carlow, Ireland
W.Burke
Kinsealy Research Centre, Dublin, Ireland

ABSTRACT

The combination of high rainfall and low evapotranspiration rates for much of the year has tended to confine the area of tillage cropping to the southern and eastern regions of Ireland. Soils tend to be wet when worked leaving them vulnerable to soil compaction. This paper outlines research findings from cultivation and deep loosening experiments for sugar beet. Some cultivation treatments have a compacting effect on the soil and reduce yields significantly. Preliminary work with autumn cultivations indicate that they facilitate earlier spring sowing but that surface crusting may be a problem leading to reduced establishment.

Deep loosening as a means of ameliorating soil compaction has given positive results where the subsequent number of tillage operations has been minimised. Beneficial effects of soil loosening are often lost through recompaction.

INTRODUCTION

The purpose of this short presentation is to set out for the Workshop the position on soil compaction in tillage soils in Ireland. It covers the role of tillage in Ireland, the main problems encountered and the research in progress to cope with these problems.

RAINFALL

Ireland lies between 53° and 55° North latitude on the Western edge of Europe in the pathway of the prevailing S.W. moisture laden winds from the Atlantic. Mountains about 1000 m high lie along most of the west coast. When the moisture laden air reaches those mountains it is forced upwards and cooled. This includes precipitation which is heaviest near the west coast and decreases in an eastward direction. A generalised map of rainfall is shown in Fig. 1. The general outline of the 1200 mm, 1000 mm and

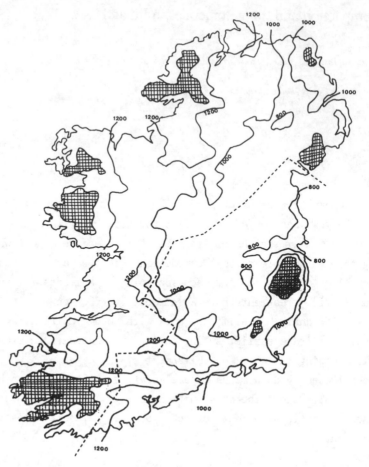

Fig. 1.   Outline of mean annual rainfall (mm), 1951-'80 and location of
           main tillage area (east of broken line) in Ireland. Shaded areas
           represent mountains (> 1600 mm mean annual rainfall).
           Adapted from Irish Meteorological Service, Dublin Climatological
           Note No. 7., 1984.

800 mm isohyets are shown. Heavier precipitation occurs locally in and near
mountains. Rainfall is fairly uniformly distributed throughout the year.
Potential evapotranspiration is about 450 mm.

        There is very little evaporation between mid October and mid March.
By the end of March evaporation averages about 1mm/day and increases to a
maximum of 3mm/day by mid summer. A consequence of this rainfall/evap-
oration pattern is that soils are normally wet until late April or early
May and have become wet again by late August/early September. Because of
this predominantly wet climate, about 90 per cent of the agricultural land

is under grass and 10 per cent is under tillage. Tillage is rare in the wetter western region and is generally confined to the drier and warmer south east (Fig. 1). The areas under the main crops are shown in Table 1.

TABLE 1   Areas under tillage crops, 1983.

Crop	Total area (ha '000)	Crop area as % of tillage area
Barley	304	62.8
Wheat	59	12.2
Oats	22	4.5
Potatoes	32	6.6
Sugar Beet	36	7.4
Others	31	6.4

Irish Agriculture in Figures, 1983.   An Foras Taluntais, Dublin.

SOILS

All of the country has been subjected to one or more glaciations in recent geological time. As a result, all of the soils are derived from compact glacial drift and are commonly classified as Brown Earths or Grey Brown Podsolics, and fall into the textural categories of loams, silty loams or silty clay loams. Mechanical analyses of topsoil from two typical soil series are given in Table 2 (Clonroche, Paulstown).

TABLE 2   Mechanical analysis of Ap1 horizons from two typical soil series in south east Ireland.

Particle size	Per cent by weight	
	Clonroche	Paulstown
2 to 0.2 mm	28	19
0.2 to 0.02 mm	16	22
0.02 to 0.002 mm	31	34
<0.002 mm	25	25

The combination of relatively fine textures and climate renders the soils very liable to compaction under tillage and this is the main problem in tilled soils in Ireland. It is very widespread and almost impossible to avoid under Irish conditions. Plough pans occur in most tilled fields, and compaction in the cultivated layer is also common. Surface crusting also occurs. Most soils have adequate organic matter because many farmers still operate mixed livestock/tillage systems and most rotations include a grass break.

However, in recent years many of the larger farmers have adopted continuous cereals systems. These systems have lead to a rapid decrease in organic matter and very severe compaction problems unless special precautions had been taken, e.g. early sowing of winter cereals, regular deep loosening, minimum traffic and very careful spring cultivations.

RESEARCH

Against this background a relatively small research programme was commenced some years ago. Most of the research work on seedbed preparation has been conducted with the sugar beet crop with a limited amount of work on cereals, sugar beet being a crop which is very sensitive to seedbed conditions and soil structure problems.

The optimum seedbed for sugar beet is one which gives rapid germination and establishment and provides a suitable environment for extensive root growth subsequently. Under Irish conditions it is vital to get the crop established early to maximise the interception of solar energy in May and June. Generally this means getting the crop sown in March or early April. On heavier finer textured soils this target sowing date is seldom achieved because the soil moisture contents are too high. In fact sowing is often delayed on these soils until late April or early May. Even if soil conditions at the surface are suitable for working the deeper layers are still likely to be vulnerable to compaction by wheel traffic and implemen pressure.

In the course of our work comparing implements for seedbed preparation we have shown that excessive wheelings and implement passes can have a serious effect on sugar beet yield and root shape (Fortune, 1978). Implements which are capable of creating seedbeds in one pass, e.g. powered rotary cultivators, generally produced the highest root and sugar yields and the least amount of forked roots (Table 3).

TABLE 3   Sugar beet root yield and shape as affected by cultivation
treatment

Cultivation treatment	Root yield (t/ha)		Root shape - % well shaped roots	
	Flat	Drills	Flat	Drills
Disc harrow (3)* + Dutch harrow (2)	45.9	46.2	47.3	58.6
Tine Cultivator (3) + Dutch harrow (2)	50.6	49.5	53.0	53.0
Powered rotary cultivator (1) + Dutch harrow (1)	54.6	54.2	53.7	62.6

( )* No. of implement passes

Because of the delay in sowing sugar beet due to the slow drying out
of soil in spring it was felt that seedbed preparation in autumn when conditions are usually better for cultivation operations might be a more
successful way of getting the beet crop sown early. Research in the U.K.
(Spoor and Godwin, 1984; Rowse and Goodman, 1984) has shown the possible
benefit of soil preparation in autumn for earlier spring sowing without
further soil manipulation

A preliminary experiment was laid down in autumn 1983 on two sites
(light and heavy soils) at Oak Park comparing autumn prepared seedbeds with
spring cultivation and sowing. The autumn seedbeds were ridged into 56 cm
rows; it is envisaged that in practice cultivation and ridging would be
carried out in one operation leaving wheel tracks in the furrows. It was
hoped to be able to sow the autumn cultivated plots earlier in spring than
those cultivated at the normal time in spring. However, dry weather conditions early in 1984 meant that spring cultivation could be conducted much

119

TABLE 4   Effect of autumn cultivation on sugar beet establishment and yield of sugar 1983-'84.

| | Establishment | | | Yield of sugar and root shape | | | |
| | Pre-singling | | Post singling | | Root shape | | |
	Plant establishment %	Plants /ha (000's)	Plants /ha (000's)	Sugar Yield (t/ha)	No forking	Slightly forked	Badly forked
Fine autumn seedbed, ridged, spring sown.	64.2	141.1	67.5	10.75	62.7	17.8	19.5
Coarse autumn seedbed, ridged, spring sown.	61.7	140.3	67.9	10.30	56.5	19.0	24.5
Spring seedbed ridged.	52.1	118.5	62.9	10.85	57.5	25.3	17.1
Spring seedbed flat sown	50.7	115.3	64.0	10.42	58.3	20.8	20.9

earlier than normal and both treatments were cultivated and sown on March 20. Some results from the heavier soil are given in Table 4. The plant establishment was 12% higher after the autumn cultivation than after the spring operations. However, the early growth was less vigorous on the autumn cultivated plots and final yields of sugar were similar (Fortune, 1985).

The experiment was repeated in 1984-'85. Preliminary results indicate that plant establishment was poorer on the autumn cultivated treatments, which were sown earlier (March 28 vs. April 20) than the spring cultivated plots. This could be attributed to considerable soil capping or crusting on the early sown treatments.

COMPACTION

The initial method adopted in relation to soil compaction was to carry out field examination in a wide variety of soils. Profiles were excavated to about 60 cm and the soils were visually examined. This was supplemented by augering and penetrometry. On selected sites deeper profiles were excavated and undisturbed samples were taken for laboratory assessments of texture, density, moisture characteristic curves and associated pore size distribution. Observations were also made on such secondary effects as surface ponding of water, forking of sugar beets and the occurrence of drought effects. It became apparent that deep-seated soil compaction was a widespread problem.

A series of experiments was initiated in 1980 to compare the effects of deep loosening to approximately 40 cm with conventional tillage in sugar beet fields. The experiment continued for four years. Various types of subsoilers were used, e.g. conventional rigid tines, winged tines, and vibrating tines. Subsoiling was carried out in dry soil conditions after cereal harvest in autumn and this was followed by conventional ploughing and seedbed preparation in springs 1981 and 1982. Some variations were made in tillage methods in 1983 and 1984.

Detailed laboratory examinations, as outlined above, were carried out on soil samples taken from all sites before subsoiling. Cone penetrometer measurements were conducted after subsoiling, ploughing and seedbed preparation. All plots were assessed for yield of roots, yield of sugar and numbers of deformed or forked roots.

During 1981 and 1982 experiments were laid down in 12 sites. Yields of roots and sugar were significantly increased on two sites only - the maximum increase recorded being 20% on a sandy loam. A significant

121

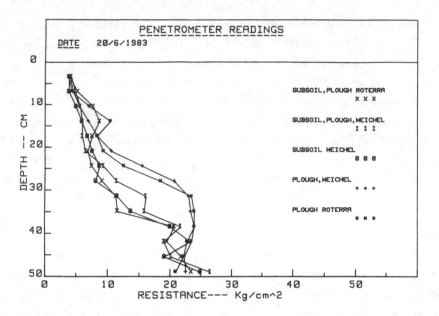

Fig. 2. Penetration resistance under various cultivation and subsoiling
treatments, 1983 site.

reduction in forking was recorded on 3 sites, 33% reduction on 2 sites and
25% on the third.

Observation in 1981 and 1982 indicated that some re-compaction of
subsoiled plots occurred during conventional spring cultivations. Accord-
ingly some different cultivation treatments were introduced during 1983 and
1984. For the 1983 experiment subsoiling was done in two directions at
right angles to achieve thorough loosening. The non subsoiled plots were
ploughed in October to 25 cm depth; some subsoiled plots were not ploughed.
Conventional type tillage or reduced tillage treatments were subsequently
superimposed on all plots. Experiments were carried out at 4 sites. Soil
between 25-35 cm depth was effectively loosened by subsoiling. Fig. 2
illustrates this effect for one of the 1983 sites.

Significant yield differences due to treatments were not recorded at
any site (Burke and Jelley, 1984). There was a significant reduction in
forking on all sites due to both subsoiling and reduced cultivation treatments (Table 5). The combined effect of subsoiling and reduced tillage
was to reduce forking by an average of more than 50%.

TABLE 5   Effect of subsoiling and cultivation treatments on root shape, 1983.

| Treatment | % Forking | | |
	Site 1	Site 2	Mean
Conventional ploughing and harrowing	37	32	35
Ploughing and reduced tillage	29	26	28
Subsoiling and conventional tillage	33	25	29
Subsoiling, ploughing and reduced tillage	23	14	18
Subsoiling, no ploughing, reduced tillage	16	13	14

In 1984 only two sites were under study. Treatments consisted of double and single subsoiling combined with conventional and reduced cultivations. On one site there was no significant effect on yield or forking while on the second site there was a 10 to 20% increase in yield due to subsoiling treatments but no effect in forking.

We attribute the variable results obtained during the four years to two main factors:-

1. The re-compaction effects of spring tillage on wet soils following autumn loosening.

2. The pattern of summer rainfall which is often adequate to maintain growth unchecked even when root depth may be restricted.

In a separate set of experiments on deep loosening and seedbed cultvation for sugar beet. Larney and Fortune (1985) found that beneficial effects of deep loosening were cancelled by subsequent tillage operations such as ploughing and seedbed preparation. Loosened soil was recompacted back to its original density. Recompaction was caused by ploughing in wet conditions, unnecessary wheel traffic on ploughed soil and over- cultivation with compacting implements. This was reflected in the low yield response of sugar beet to deep loosening. It was found that the wetter state of deep loosened soils compared with non-loosened soils, in spring, accentuated the problem of recompaction.

Larney (1985) reported on soil water relations of deep loosened and non loosened treatments and sugar beet growth using a neutron probe. In the dry growing season of 1984 deep loosening resulted in a greater number of days when soil volumetric moisture content was below permanent wilting

123

point at successive 10 cm soil depths. This was due to the extraction of greater amounts of water from deep loosened soil due to an increase in rooting density. This in turn led to an increase in root yield.

CONCLUSIONS

Our work has shown that there is a serious problem of soil structure degradation and soil compaction in Ireland. We can improve soil conditions considerably by deep loosening and by using suitable tillage methods. It is not obvious at this stage how we can preserve these improvements and turn them into increased yields. The work is continuing but in a modified form based on our experiences and more attention is being paid to the precise measurement of changes in soil structure.

REFERENCES

Burke, W. and Jelley, R.M. 1984. Sugar beet growth and soil compaction. Annual Report on Sugar Beet Research Programme, 1983. An Foras Taluntais, 29-33.

Fortune, R.A. 1979. Sugar beet cultivations. Annual Report on Sugar Beet Research Programme, 1978. An Foras Taluntais, 36-51.

Fortune, R.A. 1985. Sugar beet cultivations. Annual Report on Sugar Beet Research Programme, 1984. An Foras Taluntais. 23-29.

Larney, F.J. 1985. Effects of soil physical conditions, deep loosening and seedbed cultivations on growth and yield of sugar beet. Unpublished Ph. D thesis, National University of Ireland, Dublin.

Larney, F.J. and Fortune, R.A. 1985. Characterisation of the recompaction effects of mouldboard ploughing and seedbed cultivations on deep loosened soils. Proc. 10th Conf. Int. Soil Tillage Res. Orgn. (Guelph, Canada). In press.

Rowse, H.R. and Goodman, D., 1984. Drilling vegetables into autumn-cultivated soil with a low ground pressure vehicle: effects on timeliness and soil compaction. J. of Soil Sci., 35, 347-335

Spoor, G. and Godwin, R.J., 1984. Influence of autumn tilth on soil physical conditions and timing of crop establishment. J. agric. Engng. Res., 29, 265-273.

Some effects of deep cultivation of sandy loam soils

J. Alblas

Research Station for Arable Farming and Field Production of Vegetables, Lelystad, Netherlands

INTRODUCTION

The present study was started because a considerable difference was observed between the yields of potatoes on the "old" land in the south-west of The Netherlands and on the "new" reclaimed land. In general the plough-pan is more dense in the "old" soils than in the "new" one. But also the depth of root penetration is less in the old soils (60 cm) than in the new soils (>80 cm).

In 1978 two field trials were started to study effects of deep cultivation in sandy loam soils in the south-west of The Netherlands. This research was started to answer two questions:
- what influences do the deep cultivations have on the growth and yields of the crops;
- for how many years are the effects of deep cultivations measurable.

Both trials are situated in polders which were reclaimed about 1600 AD and are calcareous marine soils. The topsoil of the trial at Colijnsplaat is 32 cm thick and has a clay content of 13% and 2% organic matter. Below the Ap the clay content gradually decreases and little organic matter is present. Below the topsoil the soil is moderately to severely compacted. From 50 to 90 cm the soil has very fine pores, the groundwater level is more than 150 cm in summer and in the winter the highest measured level is ~40 cm. The subsoil is poorly explored by roots but water is sufficiently available. At Westmaas the topsoil, with a depth of 35 cm, contains 20% clay and 2% organic matter. Deeper down the clay content decreases to 10% at 100 cm below the surface. Generally the groundwater level is deeper than 140 cm in summer and the highest level in winter is about 60 cm. Water is sufficiently available and the subsoil is poorly explored by roots of potato and moderately for other crops. In both soils the capillary water transport is about 2 mm/24 hours.

In the trials the following treatments have been carried out under good conditions in September 1978:

125

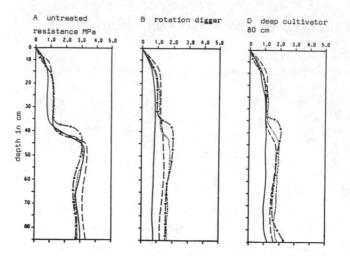

Fig. 1   Penetration resistance (MPa), Colijnsplaat.
1979 ——— 1980 ---- 1981 ···· 1982 ·—·—·

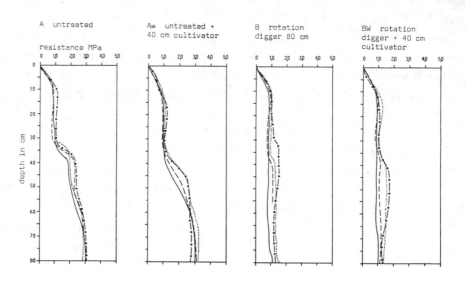

Fig. 2   Penetration resistance (MPa), Westmaas.
1979 ——— 1980 ---- 1981 ···· 1982 ·—·—·

		Colijnsplaat	Westmaas
A	untreated, no cultivation under the topsoil	+	+
B	rota digger to 80 cm	+	+
D	cultivation to 80 cm with heavy tine cultivator	+	
W	cultivated to ¯40 cm		+

126

Treatment W was done again after the harvest of cereals. There were no replications. Yearly potatoes, winter wheat, sugarbeet and barley have been grown on both trials. The following observations were made: collapsing of the surface in early spring, penetration resistance in spring (pF 2- situation), pore space, rooting depth in summer, crop growth and crop yield.

THE RESULTS

The physical situation

In the spring of 1982, almost four years after the start, influences of the various treatments are identified in the pore space (Table 1). In the layer under the topsoil - 30-35 cm - the level of the pore space of all treatments is larger than in the untreated soil. However, in the layer 40-45 cm in the treated soils the pore space slightly decreased.

TABLE 1   Pore space % below the topsoil spring 1982

		Colijnsplaat		Westmaas	
		30-35 cm	45-50 cm	30-35 cm	40-50 cm
A	untreated	40.1	47.6	40.0	44.4
AW	untreated + 40 cm cultivation	42.6	46.3	42.9	44.1
B	rotation digger 80 cm	41.6	46.8	43.2	43.7
D	depth cultivator 80 cm	42.2	45.2	–	–

During the years the treated soils again compacted as was shown by the penetration resistance. At the Colijnsplaat plot (Fig. 1) the resistance strongly increased between the first and second growing season. In this winter period a high groundwater level has been registered. At Westmaas (Fig. 2) the increase occurred more gradually. After breaking of the renewed compacted soil below the topsoil a higher penetration resistance had been found at a depth of 45 cm.

The roots

The rooting depth of potatoes increased after the cultivations done at Westmaas (Table 2) but this depended on the circumstances. The precipitation during the root growth period especially influenced the rooting depth. The more it rained in this period the smaller was the difference between the rooting depth in the untreated and treated soils.

127

TABLE 2   Depth on which 90% of the roots have been found (cm) and the penetration resistance (pr) on the same depth.   Average of 1979–1982.

| | Colijnsplaat | | | | | | Westmaas | | | |
| | A–untr. | | B–rot.digger | | D–cult. | | A–untr. | | B–rot.digger | |
	pr	cm	pr	cm	pr	cm	pr	cm	pr	cm
winter wheat	2.2	56	1.1	73	1.3	68	2.9	77	1.1	76
barley	3.3	48	1.3	68	1.1	66	3.1	73	1.0	79
potatoes	1.2	31	1.2	30	1.2	29	1.2	33	1.2	47
sugar beet	1.9	38	1.2	71	1.6	52	2.1	53	1.3	76

TABLE 3   The averages of crop yields in relative figures (A untreated = 100). Colijnsplaat – sandy loam soil.

		B – 80 cm rota digger	D – 80 cm deep cultivator
winter wheat	1980–1983	98	98
barley	1980–1981	90	96
potatoes	1979, 1981–1983	92	102
sugar beet	1979, 1980, 1982, 1983	95	99

At the trial at Colijnsplaat the potato roots did not grow into the cultivated subsoils. However, winter wheat, barley and sugar beet rooted deeper in the deep cultivated layer. At Westmaas the increase of winter wheat roots was limited because the untreated subsoil could possibly be rooted. The penetration resistance measured in both sites in the various crops showed large differences in the untreated profiles. But these differences were absent in the treated plots. This clustering of resistances shows that other limiting causes were present, possibly the air content.

The yields

In the first and second year after the start of these trials potatoes and barley showed a clear reaction on the treated fields during the crop growth. Later on only barley showed a reaction. The effects of the subsoil cultivations on Colijnsplaat (Table 3) were slight. After use of the rota digger the yields of all crops decreased. Only the potato yields did not decrease on the deep cultivator site.

128

TABLE 4   The averages of crop yields in relative figures (A untreated = 100). Westmaas - loam soil.

		B - 80 cm rot.digger	80 cm rot.digger + 40 cultivator	AW - 40 cm cultivator
winter wheat	1979–1984	99	99	100
barley	1979–1984	96	94	98
potatoes	1979–1984 (marketable)	105	105	102
sugar beet	1980	105	-	100
brussel sprouts 1981		102	-	-

On the Westmaas trial (Table 4) potato yields reacted positively on the cultivations, especially the marketable yield. Barley yields decreased on all treatments (like at Colijnsplaat). Winter wheat scarcely reacted.

These crop reactions show that a clear relationship between rooting depth and crop yields is missing. Besides this the crop reactions show that the deep cultivations carried out in 1978 do not appear to have had positive effects as far as the financial results are concerned.

CONCLUSIONS

After breaking of the ploughpan and after deep cultivation a higher pore space and lower penetration resistance were found. However, in time the soil compacted again below the topsoil and the penetration resistance of the subsoil increased.

Generally the rooting depth increased after deep cultivation on the soil where a deep penetration of the soil before treatment was not possible.

The crops reacted differently. At the sandy loam soil of Colijnsplaat the crop reaction was negative. On the loam soil of Westmaas the average reaction was neutral.

The eighth year after the start (1986) will be the last year and the cooperating institutes will now analyse the results to gain information on the duration of the effects of deep cultivation.

Conséquences de l'état du profil cultural sur l'enracinement: Cas du maïs

F.Tardieu
INRA, Laboratoire d'Agronomie annexé à la Chaire d'Agronomie, INA-PG, Thiverval-Grignon, France
H.Manichon
INA-PG, Chaire d'Agronomie, Thiverval-Grignon, France

RESUME
La pénétration des racines dans un milieu dépend de la porosité et du potentiel hydrique de celui-ci. Or ces propriétés sont spatialement très variables à un instant donné dans un sol travaillé, en raison du caractère discontinu des tassements agricoles. Il s'en suit que l'état structural de la couche travaillée joue un rôle considérable sur la disposition spatiale des racines. D'une expérimentation conduite depuis 1982 sur Maïs, il ressort que, les racines situées dans la couche labourée sont groupées dans les zones de faible résistance mécanique à la pénétration ; ce groupement est d'autant plus marqué que les obstacles structuraux sont fréquents. De plus, les obstacles de dimension décimétrique (tassements de roues, par exemple) provoquent de fortes irrégularités de colonisation dans les couches non affectées par les tassements : les volumes de sol situés à leur verticale sont significativement moins colonisés que ceux situés sous les zones enracinées de la couche labourée. D'autre part, des écarts de masse racinaire totale et de profondeur d'enracinement ont été constatés à la floraison entre des placettes d'états structuraux différenciés. Ils sont vraisemblablement liés aux différences de croissance des peuplements végétaux (partie aérienne et souterraine), induites par l'effet de la disposition spatiale des racines sur l'alimentation hydrique et minérale.

Root penetration depends on porosity and water potential of soil. In cultivated soils, those properties show a marked spatial variability, the intensity of which depends on soil structure. As a consequence, structure of the ploughed layer has a high effect on root spatial arrangement. From an experimentation carried out for 1982 on maize, it appears that in the ploughed layer, roots are concentrated in the low mechanical resistance zones. Observed patterns of root impacts (on vertical or horizontal planes) werethe most clustered when structural obstacles were frequent in the ploughed layer. Decimetrical obstacles in this layer caused also marked irregularities in the colonization of subsequent layers, which were not affected by compactions. Volumes of soil located underneath these obstacles were significantly less colonized than those located below the colonized parts of the ploughed layer. Differences in root weight were observed at flowering between treatments (defined by soil structure of the ploughed layer). They are probably due to growth differences (roots and shoots) between treatments, induced by the effect of spatial arrangement of roots on water and mineral elements uptake.

INTRODUCTION

Dans la littérature, les relations entre l'état structural du sol et l'enracinement ont souvent été étudiées à travers la modélisation de la pénétration des racines dans des milieux homogènes, en fonction des caracté-

131

ristiques physiques de ceux-ci. Des travaux effectués sur matériaux artificiels (Wiersum, 1957 ; Aubertin et Kardos, 1965) montrent les effets respectifs de la rigidité et de la porosité du milieu, ce qui est confirmé par
les études réalisées en laboratoire sur des matériaux naturels (Taylor et
Gardner, 1963 ; Barley, 1963 ; Maertens, 1964). Celles-ci montrent que la
pénétration d'une aiguille métallique, reliée statistiquement à la pénétration des racines, dépend de la porosité et du potentiel hydrique matriciel
du milieu considéré. Plusieurs modèles de pénétration des racines ont été
établis à partir de ces données expérimentales (Barley et Greacen, 1967 ;
Hewitt et Dexter, 1979, en particulier).

    Cependant, il serait dangereux de transposer sans précaution ces ré-
sultats au champ, en cherchant par exemple à établir des relations prévi-
sionnelles entre des mesures physiques effectuées au semis (porosité moyenne ou résistance moyenne à la pénétration) et l'état final du système racinaire (Raghavan et al, 1979, par exemple). En effet, au champ :
. Les caractéristiques du sol sont variables au cours du temps (variations
  d'humidité, fissuration en particulier). Les différentes générations de
  racines (Picard et al, 1985) ne rencontrent donc pas les mêmes conditions
  physiques lors de leur croissance.
. Les variations de porosité liées aux tassements agricoles ne concernent
  pas l'ensemble du système étudié, comme dans les expérimentations citées.
  Elles n'ont lieu que dans l'horizon labouré et éventuellement dans le premier horizon sous-jacent, alors que la profondeur totale d'enracinement
  est généralement supérieure au mètre. D'autre part, elles ne concernent
  que la partie de l'horizon labouré située sous les roues des engins. Il
  s'en suit une grande variabilité horizontale de la porosité, entre zones
  distantes de quelques centimètres (Manichon, 1982). Sauf cas assez rares,
  d'obstacles continus à la pénétration des racines, il existe dans la qua-
  si totalité des profils des zones de faible résistance mécanique à la pénétration (fentes de retrait, inter-bandes, zones non tassées) permettant
  le passage des racines vers des zones non affectées par le tassement.
  L'étude de l'effet des tassements ne se ramène donc pas à l'application
  des modèles de pénétration des racines dans un milieu caractérisé par sa
  densité apparente moyenne. Elle a plus trait à l'effet de l'état structural de la couche labourée sur la répartition spatiale des racines dans le
  volume colonisé.
. Enfin, les expérimentations citées sont effectuées au cours de temps
  courts, sur des plantules. Elles n'intègrent donc pas l'effet de l'état
  structural sur la croissance des plantes (mise en évidence par de nombreux

auteurs : Manichon, non publié ; Raghavan et al, 1979 ; Tardieu, 1984),
qui se répercute sur l'état de croissance du système racinaire.

L'article présenté ici a pour but de synthétiser des résultats obtenus de 1981 à 1984 sur les relations entre l'état structural, la répartition spatiale des racines et la croissance des systèmes racinaires.

MATERIEL ET METHODES

Dispositif expérimental

Le dispositif a été mis en place à GRIGNON (Yvelines) depuis 1982,
ainsi qu'à MONTMIRAIL (Marne) en 1982, sur des sols de limons profonds. Les
traitements expérimentaux sont constitués par des états structuraux différenciés de la couche labourée. Trois états-types ont été définis :
. Le premier, noté "O", correspond à un état structural fortement fragmenté
  ne comportant pas d'obstacles de grandes dimensions, ni de cavités.
. Le second "B", comporte au contraire des blocs de plusieurs décimètres-
  cubes, fortement tassés et séparés par des cavités.
. Le troisième "C", correspond à un état continu, fortement compacté.

Ces trois états ont été obtenus par des itinéraires techniques appro-
priés : labour d'hiver en O ; tassement intense et homogène, puis labour de
printemps en conditions humides en B ; tassement suivant le labour en C.
Les placettes d'observation des racines ont été disposées sur la parcelle
afin que les roues des tracteurs réalisant les façons superficielles ne per-
turbent pas les états ainsi créés (Tardieu, 1984).

En 1984, un état supplémentaire a été créé, noté "A". Il est constitué d'un état de Type O sur deux tiers du profil (inter-rang et rang de semis) et d'un tiers de Type C (inter-rang tassé).

Ces traitements ont été placés sur un dispositif à trois blocs, sans
répétition intra-bloc. Le maïs (variété LG 1) a été semé début mai ; le
désherbage et la fertilisation ont été effectués suivant les normes régio-
nales.

Contrôle de l'état structural

Une fosse a été ouverte dans chaque parcelle expérimentale immédiatement après semis pour vérifier que les différenciations souhaitées avaient
été obtenues. L'état des horizons a été caractérisé d'une part par des mesures de densité (sonde à atténuation γ), d'autre part à l'aide de la méthode d'observation proposée par Manichon (1982). Celle-ci consiste à distinguer, au sein de chaque horizon, des unités morphologiques dont l'état
structural est défini, d'une part par l'assemblage des éléments structuraux
(états massifs "M", ou fragmentaires "F"), d'autre part par l'état interne

133

		Pourcentage de surface du profil occupé par les différents états structuraux				État interne dominant des mottes	Densité apparente moyenne de l'horizon	
	Massif	Massif fissuré	F grosses mottes+ cavités	F petites mottes+ cavités	F terre fine +petites mottes		$\overline{x}$	i
TRAITEMENT O								
. GRIGNON 1983	13	4	13	13	57	Δ	1.30	0.03
. MONTMIRAIL 1982	24	5	15	9	47	Δ	1.45	0.01
. GRIGNON 1982	10	0	0	8	82	Γ	1.41	0.02
TRAITEMENT B								
. GRIGNON 1983	43	7	38	2	10	Δ	-	-
. MONTMIRAIL 1982	37	7	50	0	6	Δ	-	-
. GRIGNON 1982	34	8	47	0	11	Δ	1.43	0.01
TRAITEMENT C								
. GRIGNON 1983	100	0	0	0	0	Δ	1.69	0.02
. MONTMIRAIL 1982	95	5	0	0	0	Δ	1.63	0.01
. GRIGNON 1982	98	2	0	0	0	Γ	1.55	0.01

Tableau 1. Caractéristiques de la couche labourée.
Ploughed layer characteristics.

de ceux-ci (état "Δ" de porosité structurale très réduite, ou " Γ" plus poreux) (cf. Fig. 3).

Les états structuraux obtenus sont présentés ailleurs (Tardieu, 1984). Leurs principales caractéristiques sont :

. Seule la couche labourée (jusqu'à 28 cm) a été affectée par les tasse- ments : la densité apparente mesurée entre 30 et 50 cm de profondeur n'est pas significativement différente entre traitements d'un site (les couches situées en-dessous de l'horizon travaillé ne présentent pas d'obs- tacles à l'enracinement jusqu'à 1 m). D'autre part, les dix premiers cen- timètres de sol ont été affinés dans tous les traitements et avaient des caractéristiques favorables pour la germination et l'enracinement.

. En chacun des sites expérimentaux, un contraste apparaît entre traite- ments (Tab. 1) : les différenciations souhaitées ont pu être obtenues grâce aux itinéraires techniques utilisés. D'autre part, une grande res- semblance existe entre les états obtenus, quel que soit le site.

. Alors que les états O et B ont peu évolué au cours de la période de crois- sance racinaire (Levée-Floraison), des fentes de retrait sub-verticales sont apparues dans la couche labourée du traitement C, délimitant des élé- ments structuraux massifs polyédriques, de dimension décimétrique.

. La variabilité d'états structuraux entre placettes d'un traitement appa- raît faible par rapport aux différences inter-traitements. Une analyse

factorielle des correspondances montre l'existence de groupes disjoints
de placettes sur le critère de la proportion de surface des profils occu-
pés par les différents types d'états structuraux.

. L'humidité du sol a été suivie en 1983, au cours de la période de croissance racinaire. Il en ressort que les zones massives du profil (de faible porosité) ont vu leur humidité varier très lentement (de 20 à 18%
d'humidité pondérale entre la levée et la floraison), alors que de fortes
fluctuations ont été constatées dans les zones fragmentaires (de 23 à 15%
d'humidité pondérale).

. L'état A recherché en 1984 a également été obtenu : un tiers environ du
profil a un état structural de type massif Δ , le reste était fragmentaire, proche de celui observé en O.

Observation de l'enracinement

Celui-ci a été étudié sur des placettes de 30 x 80 cm, comprenant
deux plantes chacune (40 x 80 cm, en 1984). A chaque date d'observation,
les parties aériennes ont été coupées, séchées et pesées, et le système racinaire a été caractérisé par une méthode in situ (Tardieu, 1984). Celle-ci
consiste à cartographier les impacts de racines sur un plan vertical et
cinq plans horizontaux superposés, coupant le volume de sol situé sous la
placette. Nous nous sommes aidés pour cela, en 1982 et 1983, d'une grille de
maille carré 2 cm (une note de densité racinaire a été attribuée à chacune
des cases de celle-ci) ; en 1984, les impacts de racines ont été cartogra-
phiés directement, après application d'une feuille de transparent sur le
plan étudié. Cette méthode permet, d'une part d'évaluer pour chaque placette la profondeur d'enracinement et la masse racinaire totale (obtenue par
une régression entre notes de densité et poids sec de racines, Tardieu et
Manichon, 1985) ; d'autre part d'étudier la répartition verticale et horizontale des racines. Les cartes racinaires ont été effectuées sur les mêmes
plans où l'état structural avait été cartographié, ce qui permet une mise
en correspondance directe entre densité racinaire et état structural (Fig.3).

RESULTATS ET DISCUSSION

Masse racinaire totale

La Fig. 1 présente l'évolution de la masse racinaire sur chaque traitement pendant l'année 1983. Les courbes obtenues en O et B ont l'aspect
classique des courbes de croissance : leur forme et les niveaux atteints
sont comparables à ceux obtenus par Foth (1962) et par Mengel et Barber
(1974). La croissance, assez lente en début de cycle, s'accélère à partir

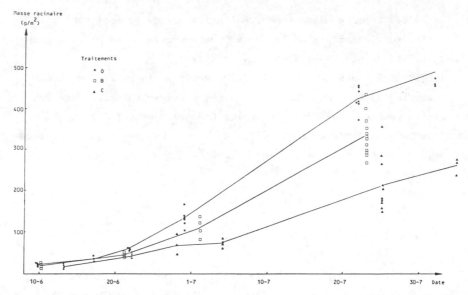

Fig.1 Croissance du système racinaire, Grignon 1983
Root system growth, grignon 1983

de la montaison, moment où l'augmentation de poids des parties aériennes devient importante. Une différenciation entre traitements apparaît à partir de ce stade, le classement était O > B > C. Dans les deux autres essais, le classement des masses racinaires à la floraison est le même que celui observé ici, et les différences sont également significatives.

L'état structural de la couche labourée a donc provoqué des différences de masse racinaire totale, bien que cette couche ne représente qu'une faible part du volume colonisé par les racines (horizon 10-28 cm). Les classements entre traitements sont les mêmes pour la masse des parties aériennes que pour les parties souterraines. La Fig. 1 reflète donc une différence de croissance entre les trois peuplements végétaux étudiés.

Répartition verticale de la densité racinaire

Celle-ci a été caractérisée sur les profils verticaux par la proportion de cases de 2 x 2 cm, où une racine au moins a été observée. La relation entre la densité racinaire ainsi évaluée et la profondeur sont présentées à la Fig. 2, pour l'année 1983. Les courbes correspondant aux traitements O et B ont des formes similaires : la densité croît avec la profondeur jusqu'à 10 cm, puis reste assez stable dans la couche labourée, avant de décroître quasi-linéairement dans les couches sous-jacentes. Les valeurs observées de densité sont cependant plus élevées en O qu'en B, tant dans la couche labourée que dans les couches sous-jacentes.

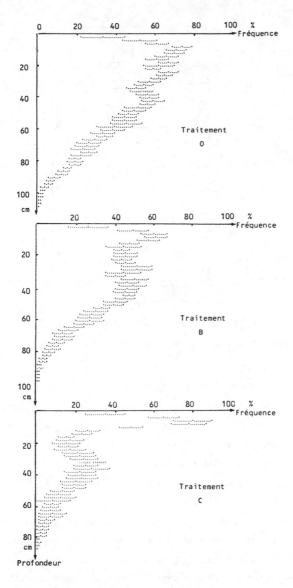

Fig.2 Densité racinaire et profondeur:moyennes et intervalles de
confiance de la proportion de cases(2x2cm) où un impact de
racine au moins a été observé.

Root density and depth:average and interval of confidence
of the proportion of 2x2cm squares where at least one root
impact was observed.

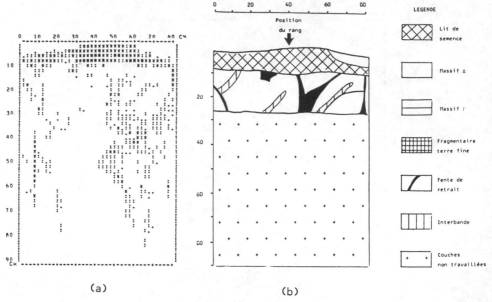

(a)   (b)

Fig.3 Cartes racinaire (a) et structurale (b) ,traitement C.
      Root(a) and stucture (b) mappings,treatment C.

En revanche, les courbes ont une forme différente dans le traitement
C : on observe une forte densité de racines dans les dix premiers centimètres de profondeur où le sol a été ameubli par le travail superficiel. A
la limite de cet horizon se produit une diminution brutale de la densité racinaire, qui reste ensuite stable dans la couche labourée. La colonisation
des couches sous-jacentes, bien que légèrement plus intense que celle de la
couche labourée, est très inférieure à celle constatée en O et B.

Les résultats obtenus sur les autres situations expérimentales étant
similaires, il apparaît que l'état structural de la couche labourée joue un
rôle sur la colonisation non seulement dans cet horizon, mais aussi dans les
horizons sus et sous-jacents : on constate, dans les quatre situations étudiées, une densité plus faible à toutes les profondeurs, en B par rapport à
O ; en C, cette réduction de densité racinaire s'accompagne d'une nette modification de la distribution verticale des racines.

Répartition horizontale dans la couche labourée

Les cartes d'impacts racinaires obtenus ont été superposées aux cartes
structurales effectuées sur les mêmes plans. Nous avons ainsi pu calculer,
dans chacun des types d'états structuraux, la fréquence de cases où une racine au moins a été observée.

138

	Types d'états structuraux						
	Massif Δ	Massif Γ	Massif Δ fissuré	F terre fine	F mottes + cavités	Fentes de retrait	χ^2 calculé
Traitement C							
Effectif (cases)	1104	0	304	0	0	246	722 S**
Proportion	8		46			88	
Traitement B							
Effectifs	426	0	301	0	799	0	632 S**
Proportion	7		66		81		
Traitement O							
Effectifs	106	359	0	665	411	0	162 S**
Proportion	17	48		71	72		

Tableau 2 Effet de l'état structural sur la localisation des racines dans la couche labourée:proportion de cases(2x2cm) où une racine au moins a été observée.

Effect of soil structure on root localization in the ploughed layer :proportion of 2x2cm squares where at least one root impact was observed.

Les résultats (Tab. 2) montrent que, dans les trois traitements, les impacts de racines sont regroupés dans les zones fragmentaires (en O et B) ou dans les fentes de retrait (en C). Le test χ^2, effectué sur les données montre que le regroupement est significatif dans tous les traitements ; il est cependant plus prononcé en B et C qu'en O.

Ces résultats, cohérents avec les études évoquées en introduction, confirment l'existence d'une forte variabilité horizontale de la densité racinaire, d'échelle décimétrique, dans la couche labourée. Celle-ci, d'origine structurale, est d'autant plus forte que les obstacles à la pénétration des racines sont plus fréquents.

Effet de l'état structural de la couche labourée sur la localisation des des racines dans les couches non travaillées

L'observation de certaines cartes racinaires verticales (Fig. 3) suggère l'idée que des obstacles de grande dimension, situés dans la couche labourée, peuvent provoquer des hétérogénéités de colonisation dans les couches sous-jacentes. Pour la vérifier, nous avons mis en correspondance les cartographies racinaires effectuées sur un plan horizontal à 17 cm de profondeur avec celles concernant les plans horizontaux sous-jacents.

a) En 1983, nous avons effectué cette étude sur trois placettes du traitement C, où la couche labourée est formée de blocs polyédriques non colonisés, séparés par des fentes de retrait sub-verticales où sont localisées les racines.

	Zonage de la couche labourée, distance à la racine la plus proche				
	< 4 (zone colonisée)	4 - 7	7 - 12	>12	χ^2 (1)
Nombre de cases	485	231	91	33	
Projection de ce zonage sur les plans situé à :					
30 cm f (%)	52	32	21	18	52 S**
60 cm f (%)	25	17	3	6	35 S**
80 cm f (%)	9	6	0	3	12 S**
100 cm f (%)	1	2	1	0	1 NS

Tableau 3   Effet de la localisation des racines dans la couche labourée sur la colonisation des couches sous-jacentes.

Effect of root localization in the ploughed layer on colonization of subsequent layers.

Pour cela, nous avons d'abord réalisé une partition du plan 17 cm, suivant l'éloignement des cases qui le constituent à l'impact de racine le plus proche. Quatre types de zones sont ainsi définis : celles situées respectivement à moins de 4 cm ; entre 4 et 7 cm ; entre 7 et 12 cm ; et à plus de 12 cm d'un impact de racine. Le zonage a ensuite été projeté sur les cartes correspondant aux plans sous-jacents des mêmes placettes. Dans chacun de ceux-ci, des zones ont ainsi été définies par leur position par rapport aux obstacles de la couche labourée. Chaque type de zone a été caractérisé par la proportion de cases où une racine au moins a été observée.

Les résultats de cette analyse (Tab. 3) montrent que sur les plans situés à 30, 60 et 80 cm de profondeur, il existe de fortes différences de densités d'impacts de racines entre les zones définies précédemment. Ces densités sont plus faibles dans les zones situées sous les parties centrales des obstacles de la couche labourée que dans celles situées sous les fentes de retrait. Ce phénomène, très net à 30 et 60 cm de profondeur, s'atténue à 80 cm et n'est pas significatif à 100 cm, profondeur à laquelle la densité de racines devient faible sur la totalité du plan.

b) En 1984, nous avons cherché à créer expérimentalement des "zones d'ombre" dans les couches non travaillées. Pour cela, nous avons eu recours à deux types d'obstacles à la pénétration des racines.
   - D'une part des obstacles structuraux constitués par les inter-rangs tassés du traitement "A".

	Rang		Inter-rang non tassé		Inter-rang tassé+obstacle		Inter-rang tassé	
Effectif	6		4		3		6	
Profondeur	$\overline{x}$	s	$\overline{x}$	s	$\overline{x}$	s	$\overline{x}$	s
17 cm	26.5	6.7	30.7	2.6	49.0	4.3	6.6	2.9
35 cm	16.2	2.6	14.8	2.3	4.8	2.4	5.9	1.5
60 cm	10.9	2.6	9.4	2.8	4.0	0.8	3.3	1.4
80 cm	4.8	2.6	5.9	1.9	1.9	0.8	0.7	0.5
100 cm	0.4	0.2	0.4	0.3	0.2	0.2	0.2	0.2

Tableau 4 Densité d'impacts de racines par dm^2 sur les plans
horizontaux, en fonction de la position par rapport au
rang de semis et aux obstacles situés dans la couche labourée

Root impact density on horizontal planes, for each position
in relation with row and obstacles

– D'autre part, des grilles de maille 0,1 mm, qui empêchent le passage
des racines de maïs (Aubertin et Kardos, 1965). Ces grilles de dimension
50 x 50 cm ont été placées horizontalement sur le fond de la couche labourée, dans un inter-rang non tassé. La pose a eu lieu quelques jours
après la levée, en creusant une tranchée aux dimensions appropriées, re-
bouchée immédiatement après avoir placé la grille. Nous nous sommes as-
surés que chacune des placettes ainsi préparées portait le même nombre
de pieds de maïs, et que la croissance de ceux-ci n'a pas été affectée
par le creusement de la tranchée.

Les cartes horizontales effectuées à la floraison sur ces placettes ont
été découpées en trois parties égales, situées soit sous le rang de semis, soit sous l'un des deux inter-rangs (ceux-ci pouvant être tassés,
non tassés avec obstacle au fond du labour ou non tassés sans obstacle).
Les densités d'impacts de racines dans chacune de ces situations sont
présentées au Tab. 4. Il apparaît que :

. les densités observées sous les rangs de semis et sous les inter-rangs
non tassés ne sont pas différentes. Ce résultat avait déjà été mis en
évidence lors d'observations antérieures (Tardieu, 1984) ;

. les densités observées sous les inter-rangs tassés sont significativement plus faibles que les précédentes. Les tassements effectués entre
les rangs ont donc provoqué une réduction localisée de la densité racinaire, sur l'ensemble de la profondeur d'enracinement ;

. les obstacles situés sur le fond du labour causent également une réduction de densité dans les zones situées à leur verticale. Les densités
observées sous les obstacles artificiels et sous les traces de roues

sont semblables, ce qui renforce l'idée que l'effet de ces dernières est lié au mode d'implantation du système racinaire, et non à des modifications structurales des couches non travaillées.

CONCLUSION

L'effet des tassements agricoles sur l'enracinement du maïs peut être décomposé, dans les situations étudiées, en quatre phénomènes d'ordres différents :

. Les zones compactées de la couche labourée ne sont pas colonisées, les racines se concentrant dans les parties du profil où la résistance mécanique à la pénétration est la plus faible.

. A cet effet direct s'ajoute une modification de la colonisation des horizons sous-jacents. Quelle que soit la nature de l'obstacle (zone tassée, située entre des fentes de retrait, traces de roue, grille située au fond de la couche labourée), on observe une forte réduction de colonisation dans les couches de sol situées à sa verticale, bien que l'état structural de celles-ci n'ait pas été modifié par les tassements.

. L'état structural de la couche labourée a donc un effet considérable sur la répartition (verticale et horizontale) de la densité racinaire dans le profil. En particulier, on observe une juxtaposition, à une même profondeur, de zones où la distance entre racines est millimétrique et de zones non colonisées où la distance à la racine la plus proche est de plusieurs centimètres. Ce type de disposition irrégulière des racines, observée dans les traitements A, B et C, entraîne (Tardieu et Manichon, à paraî-tre) une plus grande résistance au transfert d'eau entre le sol et les plantes dans ces traitements que dans le traitement O. A travers la disposition spatiale des racines, l'état structural de la couche labourée influe donc sur le fonctionnement du système racinaire en tant que capteur d'eau et, vraisemblablement, d'éléments minéraux.

. Cette différence de fonctionnement du système racinaire est une explication probable des différences de croissance entre les peuplements végétaux étudiés, différences qui concernent tant la masse des parties aériennes que la masse racinaire.

Les différences de densité racinaire entre placettes ou à l'intérieur d'un même profil ne peuvent donc pas s'interpréter seulement en termes de relations mécaniques directes : les quatre relations que nous avons évoquées, ont vraisemblablement toutes joué sur les différences d'enracinement entre traitements . Dans d'autres milieux et pour d'autres espèces que ceux que nous avons étudiés, elles seraient vraisemblablement modifiées, d'une part par les caractéristiques mécaniques des zones de profil situées sous la couche labourée, ainsi que par le poids relatif de celles-ci sur l'alimentation hydrique et minérale des plantes, d'autre part par les caractéristiques spécifiques de la plante considérée (poussée radiculaire et mode de ramification) : un travail en cours tend, en effet, à montrer que l'effet de la structure de la couche labourée s'exprime très différemment sur féverole de printemps (système racinaire pivotant) par rapport au maïs (système fasciculé).

REFERENCES

Aubertin, G.M., Kardos, L.T. 1965. Root growth through porous media under controlled conditions. I - Effect of pore size and rigidity. Soil Sci. Amer. Proc., 29, 290-293.

Barley, K.P. 1963. Influence of soil strength on growth of roots. Soil Sci., 96, 175-180.

Barley, K.P., Greacen, E.L. 1967. Mechanical resistance as a soil factor influencing the growth of roots and underground shoots. Adv. Agron., 19, 1-43.

Foth, H.D. 1962. Root and top growth of corn. Agron. J., 54, 49-52.

Hewitt, J.S., Dexter, A.R. 1979. An improved model of root growth in structured soil. Pl. and Soil, 52, 325-343.

Maertens, C. 1964. La résistance mécanique des sols à la pénétration : ses facteurs et son influence sur l'enracinement. Ann. Agron., 15, 539-554.

Manichon, H. 1982. Influence des systèmes de culture sur le profil cultural : élaboration d'une méthode de diagnostic basée sur l'observation morphologique. Thèse Docteur Ingénieur. INA-PG, Paris.

Mengel, D.B., Barber, S.A. 1974. Development and distribution of corn root system under field conditions. Agron. J., 66, 341-344.

Picard, D., Jordan, M.O., Trendel , R. 1985. Rythme d'apparition des racines primaires du maïs. I - Etude détaillée pour une variété en un lieu donné. A paraître dans Agronomie.

Raghavan, G.S., Mc Kies, E., Baxter, R., Genardon, G. 1979. Traffic-Soil-Plant (maize) relations. J. Terramechanics, 16, 69-76.

Tardieu, F. 1984. Etude au champ de l'enracinement du maïs. Influence de l'état structural sur la répartition des racines, conséquences sur l'alimentation hydrique. Thèse Docteur-Ingénieur. INA-PG, Paris.

Tardieu, F., Manichon, H. 1985. Caractérisation de la morphologie des systèmes racinaires au champ pour l'étude des relations Etat structural-Enracinement-Rendement. Colloque"Physiologie du maïs", 93-100. INRA, Paris.

Tardieu, F., Manichon, H., 1986. Caractérisation en tant que capteur d'eau de l'enracinement du maïs en parcelles cultivées. I - Discussion des critères d'étude. A paraître dans Agronomie.

Taylor, H.M., Gardner, H.R. 1963. Penetration of cotton seedling taproots as influenced by bulk density, moisture content and strength of soil. Soil Sci., 96, 153-156.

Wiersum, L.K., 1957. The relationship of the size and structural rigidity of pores to their penetration by roots. Pl. and Soil, 9, 75-85.

The specific effects of roots on the regeneration of soil structure

M.J.Goss

Soils and Plant Nutrition Department, Rothamsted Experimental Station, Harpenden, Herts, UK

ABSTRACT

Roots anchor a plant, explore the soil for mineral nutrients and water and are the site of production of several growth controlling substances. To carry out these requirements root systems grow by branching as well as by longitudinal and radial expansion. During their life roots add to the organic and inorganic materials in the surrounding soil. Some activities such as the compression and shearing of soil during root expansion are likely to reduce the stability of natural soil aggregates whereas the addition of organic matter may improve aggregate strength. The significance of different root activities is discussed in relation to the improvement of soil structural conditions. Some methods are presented for the assessment in situ of the influence of roots on soil structure. Rooting characteristics likely to encourage the improvement of a damaged structure are identified.

1. INTRODUCTION

Since we are interested in soil compaction because of its adverse effects on root growth and crop yield much of our attention has been directed towards the mechanisms underlying these responses. This has, however, lead to a greater understanding of the interactions between soil structure and root growth (Reid, 1980). In this paper the significance of the various functions of roots is reviewed for the likely influence on soil structure and its stability.

1.1 The functions of roots

Roots perform a variety of functions as far as the plant is concerned.

1.1.1 Anchorage - an obvious and very important function. At anthesis a wheat plant may have 50-100 m of roots some reaching to a depth of 2 m in well structured soils.

1.1.2 Nutrient ion uptake - plants require a wide range of cations and anions. For example potassium, calcium, iron, magnesium and phosphate are

145

all required and these are also known to be important to the structure of soils at different levels of organization.

1.1.3 Water absorption - for wheat growing in an environment with a large evaporation demand, the flux of water to unit length of root is of the order 10 μl cm^{-1} d^{-1}. In England fluxes are generally closer to 1 or 2 μl cm^{-1} d^{-1}. However, roots can over the course of the season lower soil water potential to about -2 MPa.

1.1.4 Production of growth controlling substances - a very important function for the plant but of no direct importance for soil structure. However, some of these substances appear to be concerned in the response of roots to adverse soil conditions and hence indirectly control root distribution. In addition they may also influence the activity of the microflora in the rhizosphere.

1.1.5 Release of organic and inorganic materials - both active and passive release of components into the soil occur. In general it can be regarded as a consequence of the roots fulfilling their other functions but it probably represents a very important role for soil structure development and maintenance.

1.2 Components of root growth

To have a better understanding of the effects of roots on soil structural parameters it is also important to consider their growth.

1.2.1 Elongation - maximum rates of elongation are of the order of 25 mm d^{-1} Mechanical resistance and poor aeration are likely to be prevalent in structurally damaged soils and both result in considerable slowing of extension rates. An applied pressure of only 20 kPa reduced root growth in barley by 50% (Goss, 1977). Reducing the oxygen concentration to 0.05 (v/v) had the same effect on the rate of root extension in wheat (Powell, 1985, unpublished results).

1.2.2 Radial expansion - the diameter of primary roots of most crop plants falls in the range 0.5-2 mm. The diameter of the root tip is generally about one third that of the mature root. Mechanical resistance to root extension increases radial growth by up to 30%.

146

1.2.3 Sloughing of cells - as a root grows through the soil cells of the root cap are constantly being shed. In some species epidermal cells can also be lost in the same way. Attacks by grazing soil fauna and by bacteria can also cause loss of epidermal and cortical cells.

1.2.4 Death and decay - just as some leaves and stem branches die during the season so not all roots survive. In cereal plants the maximum total root length is registered at anthesis (flowering). The death and decay of roots can occur over several months and the more lignified tissue may take several seasons before it becomes unrecognizable. Best estimates suggest that about 20% of the carbon fixed by a plant passes into the root system. As some 75% of roots are present in the Ap horizon this represents a maximum annual input into the topsoil of around 0.6 kg organic carbon per m^3 under a wheat crop, not taking account of the carbon respired by the roots. Most agricultural soils contain 10-20 kg organic carbon per m^3 so the material added from plant roots is a very small proportion of this total.

2.   POTENTIAL CHANGES IN SOILS CONSEQUENT ON ROOT GROWTH AND FUNCTION

As root grow and carry out their functions as outlined above they affect the soil through enmeshment, compression and shear, shrinkage and dehydration, alteration of charge density and particle bonding, and increase in the organic matter content.

2.1 Enmeshment

Enmeshment is a function of the extension of roots and the formation of root hairs leading to new improved anchorage of the plant and greater capacity to exploit water and nutrient resources. Structure in the soil can be modified or created since the root mat can provide a mechanical support for the soil matrix. The effectiveness of the roots is dependent on them limiting the movement of particles or aggregates and so restricting mobilization by wind and rain. Marram grass (Psamma arenaria) provides an extreme example, being a primary colonizer of sand dunes.

In soils where mycorrhizas can form, the fungal hyphae associated with roots may impart some degree of stability in the same way. Stevenson and White (1941) assessed the stability of grassland soils in terms of the mechanical support from the roots. However, even taking account of the quantity, distribution, linear dimensions and tensile strength of the roots present there were still differences in soil stability under different grass species. Enmeshment may therefore have only limited importance in improving soil structure.

2.2 Compression and shear

Although in many soils roots tend to exploit existing pores and channels they will penetrate through the soil matrix. Although their rate of extension may be greatly slowed by mechanical resistance of the soil, roots are capable of exerting pressures of up to 2 MPa on the soil (Pfeffer, 1893). Barley, Greacen and their co-workers have shown that as a root penetrates the soil matrix the main particle movement about the root is radial resulting in compression of aggregates and reorientation of particles (resulting from shear). Greacen (1968, see Nye and Tinker, 1977) observed that in a fine sand the bulk density was 7% greater immediately around wheat roots and pea radicles increased the bulk density of both a loam and a clay soil (Greacen et al., 1968; Cockroft et al., 1969) (Table 1). In a clay soil there was also some development of crack planes due to the roots but this may have been the result of water uptake (Cockroft et al., 1969).

TABLE 1   Increase in soil bulk density around roots

Soil	Initial bulk density (g cm^{-3})	Density around roots (g cm^{-3})
Fine sand*	1.40	1.50
loam**	1.50	1.53
clay‡	1.21	1.26

*   Greacen 1968 (see text)
** Greacen et al. 1968
‡   Calculated from Cockroft et al. 1969

What are the consequences for soil structure? Rogowski and Kirkham (1962) found that whereas increments of pressure applied to a clay loam after it had been extruded did increase its water stability (as measured by wet sieving), a pressure of 6.9 MPa did not reproduce the stability of natural aggregates in the soil in its field condition. Beacher and Strickling (1955) also reported that water stability of pressure treated aggregates was less than that of field soil aggregates. The stability of pressurized or sheared soils tends to increase with time (Arya and Blake, 1972; Schweikle et al., 1974) apparently because of spontaneous reorientation of water molecules and clay particles. However, it is unclear whether the stability of treated soils can recover to the initial value. Hence it seems likely that any compression of soils by roots might tend to reduce its stability. Nevertheless, increased intimacy of contact between soil particles as a result of such compression may facilitate other processes of stabilization around roots, especially by organic materials (see section 2.5).

The channels formed by the roots may be of much greater importance for improving the structure of soil for drainage and the growth of subsequent crops. Using a non-aggregated sandy loam free from organic matter, Barley (1954) found that the permeability of soil fell after maize seeds were planted and the roots began to grow, blocking some pores and eliminating others. Later the permeability increased as root decay began. Barley did not measure permeability when decay was complete but Oades (1983, private communication) showed in a similar experiment that the root channels provided a significant number of macropores that had a greater continuity than existed originally. These channels would then provide easier access for roots of a following crop but also encourage deeper perculation of water. This would be most important where damage to soils resulted in poor structure only in a thin horizon overlying more amenable material for root exploration.

2.3 Shrinkage and dehydration

Uptake of water by roots can lead to the soil undergoing several series of wetting and drying cycles during a season. The range of these cycles will not usually be more that from 0 to -2 MPa and generally early cycles will be of smaller aptitude. Reid and Goss (1982a) showed that allowing maize roots to dry the soil to -1.5 MPa, followed by a slow rewetting to -8 KPa increased the stability of a sandy loam by almost 10%. Multiple cycles have also been shown to increase the water stability of aggregates (Sillanpää and Webber, 1961). The removal of water by roots may restrict the movement of coloidal or dissolved organic molecules within the soil effectively increasing their concentration

149

around mineral particle surfaces. This may increase the rate of absorption of stabilizing agents that would otherwise be degraded by soil microbes. However, there may be an upper limit to the number of cycles above which stability can decrease again (McHenry and Russell, 1943). Evidence from several studies (Emerson, 1959; Rogowski and Kirkham, 1962; Hofman and De Leenheer, 1975; Richardson, 1976; Tisdall et al., 1978) suggests that aggregate stability improves with successive drying cycles only in soils of poor structure such as sodic or puddled clays where drying can result in shrinkage of the clay fraction followed by formation of new bonds especially with organic materials (Baver et al., 1972). Dehydration of organic and inorganic cementing agents such as mucilages, iron and the hydrous oxides of aluminium is unlikely to be effected by roots sufficiently to influence stability. In other soils the cycling is likely to have little effect or more likely will decrease aggregate stability due to the explosion of entrapped air or the stresses and strains caused by unequal rewetting and swelling (Baver et al., 1972). Particularly in clays, shrinkage will result in cracking that will tend to disrupt compact layers and permit further root exploration. Thus there may be an improvement in structure even if not instability.

2.4 Alteration of charge density and particle bonding

Crops greatly alter the absolute and relative contents of inorganic ions in the soil through nutrient uptake by the roots. As cation and anion uptake proceeds, the ionic balance in the plant is maintained by the transfer of H^+ or HCO_3^- ions into the rhizosphere. Changes in ion concentrations can exert direct effects on the stability of soil aggregates mainly by influencing the flocculation of the clay (Baver et al., 1972) though it would be strange if a Ca-clay were deflocculated in normal field practice. In some soils ions such as calcium can concentrate in the rhizosphere if their rate of supply to the root surface by mass transfer exceeds their rate of absorption (Barber and Ozanne, 1970). But in general there will be a lowering of the concentrations of the major and minor nutrients. This could result in reduced stability of poorly buffered soils particularly those which have appreciable exchangeable sodium. Reid, Goss and Robertson (1982) compared the stability of a silt loam soil in which maize had grown with an uncropped control. Treatment of aggregates with a range of sodium chloride solutions before measuring stability resulted in greater increases in the stability of the cropped soil than in the control suggesting that cation uptake by the maize had reduced aggregate stability.

Lowering of the hydrogen ion concentration in acid soils might lead to enhanced stability if iron and aluminium hydroxides start to precipitate.

The work of Reid, Goss and Robertson (1982) also indicated that nutrient uptake could have important consequences for soil structural stability. Roots of maize reduced aggregate stability of a silt loam and a sandy loam by 12.4% and 21.6% respectively. The use of chemical pretreatment of the aggregates with acetylacetone before testing for stability showed that nearly all the loss of stability with maize could be accounted for by the destruction of organic matter/iron or aluminium/ mineral particle linkages. Since the organic matter components in these linkages was unlikely to have been effected (Reid and Goss, 1982b), removal of the iron and aluminium cations was the most likely mechanism. Evidence for the presence of chelating agents such as lactate and citrate in the rhizosphere (Matsumoto et al., 1979) encouraged speculation that these were the effective agents. In the work of Reid, Goss and Robertson the total uptake of iron plus aluminium was less than 50 mg per plant or less that 2% of the total extractable by acetylacetone. They speculated that the chelation of the iron and aluminium might render phosphate associated with the complexes more available to the plants. As the concentration of potassium and magnesium ions also increased in the soil solution as aggregates were broken down this would make these nutrients more available to the plant.

The evidence so far suggest that plant roots will tend to reduce the stability of the soil aggregates in their uptake of mineral nutrients.

2.5 Changes in soil organic matter content

There is a very considerable literature on the effect of organic matter on soil structure and stability. As roots decay they will contribute to the total organic matter but, as indicated in section 1.2.4 much of this is confined to the topsoil and with conventional cultivations may be oxidized rapidly. What is less clear is the contribution that living roots and their associated microflora make.

The total loss of organic matter from roots into the soil by secretion, leakage, autolysis and sloughing of cells has been calculated to be equivalent to 20-80% of the root dry weight (Whipps, 1984). The presence of environmental factors including solid particles (i.e. as opposed to solution culture), microorganisms (Barber and Gunn, 1974; Martin, 1977) and lowered soil water potential (Reid and Mexal, 1977) appear to encourage this loss.

The materials released include growth substances, amino acids, simple sugars, organic acids and saccharides such as hydrophilic polyglucuronate. Mycorrhizal fungal hyphae also appear to produce similar products. All these materials are released within the rhizosphere and are assimilated or modified by the many free living microbes in the root zone. In field soils it is the products of the root and the associated microbial population that are important for soil structure and stability.

Living roots of a range of species have been shown to increase the amount of materials in the soil which are readily oxidised by periodate. Generally these are polysaccharides and increase the stability of soil (Reid and Goss, 1981; Reid, Goss and Robertson, 1982; Tisdall and Oades, 1979; Baldock, 1985).

The loss of organic matter from roots into the soil also has another consequence. As the materials are more readily available to microbes than are residues from even the previous crop there is less breakdown of these materials in the presence of living roots than in fallowed soil (Reid and Goss, 1982b; Sparling, Cheshire and Mundie, 1982).

The effects of the root exudates will thus be to stabilize the structure around the roots.

3. FIELD APPLICATIONS IN AGRICULTURAL SOILS

The balance of evidence so far reviewed suggests that for the natural recovery of structure in damaged soils two aspects of root growth are important. Firstly the formation of continuous, vertically orientated pores will be of major importance both for drainage and crop growth. To this end when soils are mechanically weak (generally when they are close to field capacity) rapid root growth needs to be encouraged. For promoting drainage a few large holes per square metre would probably be adequate suggesting that fast growing strong tap rooted plants would be more appropriate. To encourage deep root growth for subsequent crops the many pores of smaller dimension such as produced by a fibrous root system would be better. If soil and plant can be matched then a crop that might start the disruption of more massive structure in the damaged soil would be useful particularly if it also encouraged reaggregation into finer units. Crops such as lucerne appear to have some of these properties.

A number of techniques are available to study the influence of roots non-destructively in the field. Changes in soil porosity can be assessed using radioisotopic techniques such as gamma-density probes or by relief metering. Less directly, moisture release characteristics can be determined in situ using a calibrated neutron moisture meter for values of volumetric water content and simultaneous readings of tensiometers for values of potential. Then the pore size distribution can be calculated. The continuity of macropores can be determined from hydraulic conductivity measurement at field saturation.

Root growth and distribution can be assessed using a variety of techniques. Water extraction from different horizons can be measured with time and the length of root determined if the volume influx of water per unit length of root per unit potential gradient between root and soils is known. Comparative measurements can be made using more direct measurement of roots growing against the walls of mini-rhizotron tubes. A light source and mirror or a video camera can be used to count the roots.

Goss et al. (1984) found a close correlation between root development and the sorptivity associated with macropores between 3 mm and 200 micrometres in diameter. This might provide a simple means of assessing improvements in soil structure following damage.

4. CONCLUSIONS

Root growth and function are likely to change soil structural parameters. Many aspects of these activities are likely to reduce the stability of soil aggregates but water uptake can create cracks and root extension can create macropores. The release of organic matter from roots will tend to stabilize soil structure. At present there is only limited information on the capability of different crops to improve damaged structure but means of assessing it are available.

ACKNOWLEDGEMENTS

I gratefully acknowledge the many discussion with J.B. Reid (Department of Soil Science, Lincoln College, Canterbury, New Zealand) when he worked with me at the Agricultural and Food Research Councils, Letcombe Laboratory and subsequently in England and New Zealand.

REFERENCES

Arya, L.M. and Blake, G.R. 1972. Stabilization of newly formed soil aggregates. Agron. J., **64**, 177-180.

Baldock, J.A. 1985. Soil aggregation and cropping systems. Thesis for degree of Master of Science, University of Guelph, Canada.

Barber, D.A. and Gunn, K.B. 1974. The effect of mechanical forces on the exudation of organic substances by the roots of cereal plants grown under sterile conditions. New Phytol., **73**, 39-45.

Barber, S.A. and Ozanne, P.G. 1970. Autoradiographic evidence for the differential effect of four plant species in altering the Ca content of the rhizosphere soil. Soil Sci. Soc. Am. Proc., **34**, 635-637.

Barley, K.P. 1954. Effects of root growth and decay on the permeability of a synthetic sandy loam. Soil Sci., **78**, 205-210.

Baver, L.D., Gardner, W.H. and Gardner, W.B. 1972. Soil Physics (Wiley and Sons, New York).

Beacher, B.F. and Strickling, E. 1955. Effect of puddling on water stability and bulk density of aggregates of certain Maryland soils. Soil Sci., **80**, 363-373.

Cockroft, B., Barley, K.P. and Greacen, E.L. 1969. The penetration of clays by fine probes and root tips. Aust. J. Soil Res., **7**, 333-348.

Emmerson, W.W. 1959. The structure of soil crumbs. J. Soil Sci., **10**, 235-244.

Goss, M.J. 1977. Effects of mechanical impedence on root growth in barley (Hordeum vulgare L.). 1. Effects on the elongation and branching of seminal root axes. J. exp. Bot., **28**, 96-111.

Goss, M.J. Ehlers, W., Boone, F.R., White, I. and Howse, K.R. 1984. Effects of soil management practice on soil physical conditions affecting root growth. J. agric. Engng. Res., **30**, 131-140.

Greacen, E.L., Farrell, D.A. and Cockroft, B. 1968. Soil resistance to metal probes and plant roots. Trans. 9th Int. Congr. Soil Sci. (Adelaide). Vol.1, 769-779.

Hofman, G. and De Leenheer, L. 1975. Influence of soil prewetting on aggregate unstability. Pedologie, **25**, 190-198.

Martin, J.K. 1977. Factors influencing the loss of organic carbon from wheat roots. Soil Biol. Biochem., **9**, 1-7.

Matsumoto, H., Okada, K. and Takahashi, E. 1979. Excretion products of maize roots from seedling to seed development stage. Plant and Soil, **53**, 17-26.

McHenry, J.R. and Russell, M.B. 1943. Elementary mechanics of aggregation of puddled materials. Soil Sci. Soc. Am. Proc., **8**, 71-78.

Nye, P.H. and Tinker, P.B. 1977. Solute transport and plant growth in soil. (Blackwell Scientific Publications, Oxford).

Pfeffer, W. 1893. Druck- und Arbeitsleistung durch Wachsende Pflanzen. Abh. sachs. Akad. Wiss., **33**, 235-474.

Reid, C.P.P. and Mexal, J.G. 1977. Water stress effects on root exudation by lodgepole pine. Soil Biol. Biochem., **9**, 417-421.

Reid, J.B. 1980. The interaction of roots and the soil with special reference to soil structural stability. Thesis for degree of Doctor of Philosophy, Reading, England.

Reid, J.B. and Goss, M.J. 1981. Effect of living roots of different plant species on aggregate stability of two arable soils. J. Soil Sci., **32**, 521-541.

Reid, J.B. and Goss, M.J. 1982a. Interactions between soil drying due to plant water use and decrease in aggregate stability caused by maize roots. J. Soil Sci., **33**, 47-53.

Reid, J.B. and Goss, M.J. 1982b. Suppression of decomposition of ^{14}C-labelled plant roots in the presence of living roots of maize and perennial ryegrass. J. Soil Sci., **33**, 387-395.

Reid, J.B., Goss, M.J. and Robertson, P.D. 1982. Relationship between the decrease in soil stability affected by the growth of maize roots and changes in organically bound iron and aluminium. J. Soil Sci., **33**, 397-410.

Richardson, S.J. 1976. Effect of artificial weathering cycles on the structural stability of a dispersed silt soil. J. Soil Sci., **27**, 287-294.

Rogowski, A.S. and Kirkham, D. 1962. Moisture, pressure and formation of water-stable soil aggregates. Soil Sci. Soc. Am. Proc., **26**, 213-216.

Schweikle, V., Blake, G.R. and Arya, L.M. 1974. Matric suction and stability changes in sheared soil. Trans. 10th Int. Congr. Soil Sci. (St. Paul), Vol.1 187-193.

Sillanpää, M. and Webber, L.R. 1961. The effect of freezing-thawing and wetting-drying cycles on soil aggregation. Can. J. Soil Sci., **41**, 182-187.

Sparling, G.P., Cheshire, M.V. and Mundie, C.M. 1982. Effect of barley plants on the decomposition of ^{14}C-labelled soil organic matter. J. Soil Sci., **33**, 89-100.

Stevenson, T.M. and White, W.J. 1941. Root fibre production of some perennial grasses. Soil Agric., **22**, 108-118.

Tisdall, J.M., Cockroft, B. and Uren, N.C. 1978. The stability of soil aggregates as affected by organic materials, microbial activity and physical disruption. Aust. J. Soil Res., **16**, 9-17.

Tisdall, J.M. and Oades, J.M. 1979. Stabilization of soil aggregates by the root system of ryegrass. Aust. J. Soil Res., **17**, 429-441.

Whipps, J.M. 1984. Environmental factors affecting the loss of carbon from the roots of wheat and barley seedlings. J. exp. Bot., **35**, 767-773.

Concluding session

I. CRITERIA TO EXPRESS COMPACTION RESPONSES IN THE SOIL

1) Pore space analysis

Changes in bulk density (or any property derived from pore and solid volume) have limitations for characterising compaction and are poorly correlated with plant growth responses.

Pore space can be partitioned into the pore space due to cultivation as well as weather conditions (structural pore space) and the pore space due to packing of soil constituents (textural pore space). The latter (and its variations with water content) can be considered as a reference state.

This partition may be convenient to aid the interpretation and to predict various physical properties.

Morphological data and fluid transfer properties should be taken into account to relate soil compaction to soil behaviour relevant for plant performance.

2) Gas diffusion

Gas permeability and diffusion need to be studied at various moisture contents. Permeability (mass-flow of air) is more sensitive than diffusion to the changes in pore structure (pore diameter in particular) that result from compaction.

Commonly, however, the inputs and outputs of compaction models are expressed only in terms of pore volume. Thus mechanical studies of soil susceptibility to compaction have required a relationship, often assumed, between pore volume and gas diffusivity. Diffusion thresholds for root functions and plant growth should be determined to evaluate limits of aeration due to compaction.

At present, measurement of relative diffusivity provides a good relative value for the comparison of physical states - including pore structure and the degree of saturation. Thus it appears there are good opportunities for using such measurements in the detailed analysis of the consequences of compaction.

The assessment of soil aeration, in terms of the risk to crop growth needs further modelling of gas exchange (O_2, CO_2) to include the consumption of oxygen by microorganisms (and plant roots).

3) Non polar-liquids retention

The interpretation of the water release characteristic curves in terms of pore size distribution is impossible (or, at least difficult) for swelling soils.

Equivalent diameters of pores derived from suction values are only the upper limits of the diameter of saturated pores. Generally this will not describe the actual organisation of particles and the pores between them. Replacing water by liquid having little or no effect on swelling or shrinkage is, perhaps, an interesting attempt to overcome some of the limitations of mercury porosimetry.

4) Morphology

Morphology change as a result of a compaction treatment and the changes can be recorded morphometrically (e.g. by image analysis).

The interest in micromorphology is well established even if the data it generates are essentially descriptive and only a little quantitative. Macromorphology, described in situ, may give a basis on which sampling can be carried out for micromorphology and the measurement of soil properties. Furthermore, macromorphology can reveal the effect of compaction which occurred prior to the last tillage and so give information about the durability of consequence in terms of particle organisation, pore size distribution and density. This is currently a real gap in risk assessment.

The results presented showed that consequences of compaction are not totally eliminated by soil tillage.

II. MODELLING

Existing models of compaction based on soil mechanics describe the stress distribution in soil under wheels, but these assume the soil to be homogeneous and isotropic medium. Such models can be readily applied for small deformations, usually found in subsoils. Topsoils, however, exibit vertical anisotropy and discontinuity of structure and hence these models are inappropriate. Nevertheless, models are needed since we cannot examine in more than a few experiments either in the field or laboratory the effects of many factors such as load, soil texture, bulk density and moisture content, each of which vary but are mutually dependent. The models are needed to integrate research findings. They are also needed to provide predictions that can be tested easily in the field or laboratory such that the general applicability of a model can be verified.

The great need is to develop theory of soil mechanics for unsaturated tilled soils.

A short-term but more empirical solution might be to elaborate simpler models to take account of the complexity of tilled layers. At least a classification of soil behaviour should then be possible.

III. REGENERATION AFTER COMPACTION

The present knowledge on clay minerals gives a guide to elaborate hypothesis about swelling and shrinkage processes and the magnitude of their variations. Care is needed in extrapolating results obtained on remoulded clay and other model soils to agricultural soil. Further laboratory, and field studies are still necessary to permit this extrapolation.

Field experiments have shown that measurements of swelling and shrinkage are not sufficient to predict cracking since the hypothesis that soil layers swell and shrink isotropically does not always hold.

The different components (horizontal and vertical) of swelling and shrinkage require assessment. Moreover, cracking is not a monotonous process either while wetting or drying. Its intensity as well as the morphology of the crack network vary according to the range of water content.

More work is required to understand and to predict cracking.

IV. SOIL COMPACTION AND ROOT SYSTEMS

Important progress has been made in understanding the interactions between soil structure and root growth. In particular, the effects of compaction on root growth and the effects of roots on soil aggregate stability.

In spite of the laborious nature of the work, results have shown the benefit of making direct observations on root systems in the field. For example, macromorphology coupled with nearest neighbour analysis of rooting patterns showed how the exploration of the subsoil by maize roots was affected by different levels of compaction in the topsoil. All the variables governing such a phenomenon still need to be defined.

An inventory of the ways roots can influence the evolution of soil structure (particularly soil stability) can be drawn up from laboratory studies. However, the relative importance of these effects and their consequences for the behaviour of arable soils remains to be quantified.

V. AGRICULTURAL CONSEQUENCES

The numerous attempts to relate soil compaction and yield have generally given conflicting results. There is a need for a more analytical approach where consequences of compaction are related to the different phases of growth and development (particularly those included in existing models for predicting crop yields). Research on root development, gas diffusion and flow in soils are good examples where analytical approaches are leading to a greater understanding of the risks associated with soil compaction.

Dr CULLETON

This workshop has tried to establish the current state of knowledge on soil compaction and regeneration. A major cause of concern is that because of their high productivity the better soils are the most likely to be damaged. In view of the surplus of certain crops in the EEC and the likelihood of major changes in land use in the future it would be useful to know which soil types are most at risk from compaction. Also, it is essential to find ways and means of preventing or, at least reducing, soil compaction either through crop rotation, better design of farm machinery or some other means.

Dr VAN WIJK

A classification of soils according to their sensitivity to compaction is, of course, necessary but it has to take account of the type of crop. The limiting factors are not necessarily the same for root-crops and cereals.

Risk assessment needs to take account of the development of contributary factors. For instance, it may be assumed that the power and weight of vehicles have not yet reached a maximum.

Models of water movement are a useful additional aid in prediction of risks because susceptibility to compaction varies with soil moisture condition. Much progress has been made in the subject.

More generally, soil physics as developed for agricultural purposes and soil mechanics as developed by civil engineers, have to be more systematically combined to elaborate sufficient compaction models, to be used in compaction risk prediction.

A major limitation to the application of soil mechanics to agricultural soils is that much of the theory was developed for saturated soils and its extension to the unsaturated state has been very limited.

Dr B.C. BALL

The experience of SIAE shows that it is possible to define physical criteria to establish soil suitability for various crops or crop systems. This suitability is mainly related to soil susceptibility to compaction.

The less suitable soils require more tillage but also more carefully controlled traffic to prevent compaction. A gas diffusion and flow model may allow definition of the suitable physical conditions. Classification of "difficult" soils gives many problems which require further study and development. Existing land use capability assessments within the EEC may possibly be adapted for identification of soils most vulnerable to compaction.

Dr ALTEMULLER

The diagnosis of soil compaction by micromorphological observations is possible. But criteria depend on texture. Soil behaviour also varies according to texture.

It seems possible, that the susceptibility of soils to compaction might be classified using micromorphology together with observations on development and distribution.

Morphology is not only descriptive.

In order to get as close as possible to the reality of physical phenomena, even in a simplified way, modelling needs some morphological basis.

Models of particle and aggregate organisation in particular must take into account the shape of the particles and not assume that they are spherical.

Moreover, the synthesis of the morphological approach and the physical approach to describing compaction is essential. Each time such a synthesis has been realised it has been very fruitful.

The need for research in this subject are so big that any element of information, even incomplete, is extremely useful. The first steps in long term projects should be encouraged. The first and most effective contribution to the application of micromorphology in these studies would be one that permitted the differences in behaviour between soil types to be analysed.

Dr PAGLIAI

In agreeing with the conclusions of Dr Altemuller it must be stressed that an efficient approach to the analysis of compaction needs the collaboration of morphologists, physicists and mineralogists. They must use quantitative methods for morphology (image analysis).

Root specialists are to be closely involved in those programmes.

Dr GOSS

Within the session on roots the two communications were concerned with contrasting aspects, the effects of compaction on rooting and the effects of roots upon soil structure regeneration.

It should be emphasised that the influence of roots on soil structure has to be studied within the range of water content where roots are active.

Concerning the interpretation of the water-release curve, it seems that, despite the critique of its use in swelling soils, the equivalent diameter may usefully be determined from the $\psi = f(\theta)$ curves within the range of very low suctions where pores are of a size similar to exploring roots.

Clearly root distribution is very significantly affected in layers other than where soil is compacted. If we are to predict how roots affect structural regeneration then care must be taken to match results of laboratory experiments to the known conditions under which roots grow. For example, roots will not dry the soil much beyond -20 MPa, so this provides an upper limit to confining stresses that roots can help generate thereby constraining the extent of fissuring in soils caused by water uptake. Similarly it provides a lower limit for the size of pores that will be drained.

List of participants

BELGIUM

-Dr L. RIXHON
Centre de Recherches Agronomi-
ques de Gembloux
Station de Phytotechnie
Chemin de Liroux
B - 5800 GEMBLOUX

E.C.C.

-D. CULLETON
Comite Land and Water Use
and Management
DG VI
Rue de la Loi 200
B - 1049 BRUXELLES

DENMARK

-P. SJØNNING
Statens Forsgsstation
Siltoftvej 2
DK - 6280 HOJER

FRANCE

-C. CHEVERRY
I.N.R.A.-E.N.S.A.
Laboratoire de Science du Sol
65, rue de St Brieuc
F - 35042 RENNES CEDEX

-P. CURMI
I.N.R.A.-E.N.S.A.-
Laboratoire de Science du Sol
65, rue de St Brieuc
F - 35042 RENNES CEDEX

-J. GUERIF
I.N.R.A. - C.R. d'Avignon
Station de Science du Sol
B.P. 91
F - 84140 MONTFAVET

-V. HALLAIRE
I.N.R.A. - C.R. d'Avignon
Station de Science du Sol
B.P. 91
F - 84140 MONTFAVET

-M. JAMAGNE
I.N.R.A.
Service d'Etude des Sols et de la
Carte Pédologique de France
C.R. d'Orléans - ARDON
F - 45160 OLIVET

-H. MANICHON
I.N.A.-P.G.
Chaire d'Agronomie
F - 78850 THIVERVAL-GRIGNON

-G. MONNIER
I.N.R.A. - C.R. d'Avignon
Station de Science du Sol
B.P. 91
F - 84140 MONTFAVET

-M. ROBERT
I.N.R.A. - C.N.R.A.
Station de Science du Sol
Route de Saint Cyr
F - 78000 VERSAILLES

-P. STENGEL
I.N.R.A. - C.R. d'Avignon
Station de Science du Sol
B.P. 91
F - 84140 MONTFAVET

-F. TARDIEU
I.N.R.A.-I.N.A.-P.G.
Laboratoire d'Agronomie
F - 78850 THIVERVAL-GRIGNON

-D. TESSIER
I.N.R.A. - C.N.R.A.
Station de Science du Sol
Route de Saint Cyr
F - 78000 VERSAILLES

F.R. of GERMANY
-Dr H.J. ALTEMULLER
Institut für Pflanzenernährung
und Bodenkunde
(FAL) Braunschweig-Völkenrode
Bundesallee 50
D - 3300 BRAUNSCHWEIG

-Dr H.G. FREDE
Institut für Bodenwissenschaften
der Universität - Göttingen
Von Sibolddstrasse 4
D - 3400 GOTTINGEN

GREECE
-S. AGGELIDES
Ministry of Agriculture,
Soil Science
Institute of Athens - Lycovrissi
GR - ATTIKI

-A. LOUISAKIS
Ministry of Agriculture,
Land Reclamation
Institute Sindos
GR - THESSALONIKI

IRELAND
-W. BURKE
Kinsealy Research Centre
Malahide Road
IRL - DUBLIN 5

-T. FORTUNE
An Foras Taluntais
Dakpark Research Centre
IRL - CARLOW

ITALY
-G. GOVI
Istituto di Agronomia Generale
Via Filippo Re 8
IT - BOLOGNE

-M. PAGLIAI
C.N.R.
Istituto per la Chimica del Terreno
Via Corridoni 78
IT - 56100 PISA

-PAOLINI
Istituto di Agronomia Generale
Université delle Tuscia
IT - VITERBO

-R. VOLTAN
Istituto di Agronomia
Via Gradenigo 6
IT - 35100 PADOVA

- M. ZILLIOTO
Istituto di Agronomia
Via Gradenigo 6
IT - 35100 PADOVA

-Dr D.B. DAVIES
MAFF, Block C, Government Buildings
Broakland Avenue
UK - CAMBRIDGE CB 2 2DR

-Dr M. GOSS
Rothamsted Experimental Station
UK - HARPENDEN, HERTS IL5 2JQ

LUXEMBOURG
-J. FRISCH
Administration des Services Techniques
de l'Agriculture
16, Route d'Esch
1019 LUXEMBOURG

NETHERLANDS
-J. ALBLAS
Proefstation voor de Akkerbouw
en de Groenteteelt in de
Vollegrond (PAG V)
P.O. Box 430
NL - 8200 AK LELYSTAD

-A.L.M. Van WIJK
Institute for Land and Water
Management Research (ICW)
P.O. Box 35
NL - 6700 AA WAGENINGEN

UNITED KINGDOM
-B.C. BALL
S.I.A.E.
UK - PENICUIK - SCOTLAND

167